Free Resolutions in
Commutative Algebra and
Algebraic Geometry

Research Notes in Mathematics

Volume 2

Research Notes in Mathematics

Volume 2

Free Resolutions in
Commutative Algebra and
Algebraic Geometry
Sundance 90

Edited by

David Eisenbud
Department of Mathematics
Brandeis University
Waltham, Massachusetts

Craig Huneke
Department of Mathematics
Purdue University
West Lafayette, Indiana

CRC Press
Taylor & Francis Group
Boca Raton London New York

CRC Press is an imprint of the
Taylor & Francis Group, an **informa** business

First published 1992 by Jones and Bartlett Publishers, Inc.

Published 2018 by CRC Press
Taylor & Francis Group
6000 Broken Sound Parkway NW, Suite 300
Boca Raton, FL 33487-2742

© 1992 by Taylor & Francis Group, LLC
CRC Press is an imprint of Taylor & Francis Group, an Informa business

No claim to original U.S. Government works

ISBN-13: 978-0-86720-285-4 (pbk)
ISBN-13: 978-1-138-45429-3 (hbk)

Visit the Taylor & Francis Web site at
http://www.taylorandfrancis.com

and the CRC Press Web site at
http://www.crcpress.com

Library of Congress Cataloging-in-Publication Data

Free resolutions in commutative algebra and algebraic geometry :
   Sundance 90 / edited by David Eisenbud and Craig Huneke.
      p.   cm. -- (Research notes in mathematics ; 2)
   ISBN 0-86720-285-8
    1. Free resolutions (Algebra)--Congresses.   2. Commutative
algebra--Congresses.   3. Geometry, Algebraic-- Congresses.
I. Eisenbud, David.   II. Huneke, C. (Craig)   III. Series : Research
notes in mathematics (Boston, Mass.) ; 2.
QA169.F72   1992
512'.24--dc20   91-45056
                                           CIP

Contents

Other Topics   91

Contributors

Numbers in parentheses refer to the pages on which the authors' contributions begin.

Luchezar L. Avramov (3), *Department of Mathematics, Purdue University, West Lafayette, Indiana 47907*

Dave Bayer (79), *Department of Mathematics, Barnard College, New York, New York 10027*

Hara Charalambous (25), *Department of Mathematics, State University of New York at Albany, Albany, New York 12222*

David Eisenbud (51), *Department of Mathematics, Brandeis University, Waltham, Massachusetts 02254*

E. G. Evans, Jr. (25), *Department of Mathematics, University of Illinois at Urbana-Champaign, Urbana, Illinois 61801*

Craig Huneke (93), *Department of Mathematics, Purdue University, West Lafayette, Indiana 47907*

Sheldon Katz (109), *Department of Mathematics, Oklahoma State University, Stillwater, Oklahoma 74078*

George R. Kempf (47), *Department of Mathematics, The Johns Hopkins University, Baltimore, Maryland 21218*

Matthew Miller (35), *Department of Mathematics, University of South Carolina, Columbia, South Carolina 29208*

Paul C. Roberts (121), *Department of Mathematics, University of Utah, Salt Lake City, Utah 84112*

Mike Stillman (79), *Department of Mathematics, Cornell University, Ithaca, New York 14853*

Bernd Ulrich (133), *Department of Mathematics, Michigan State University, East Lansing, Michigan 48824*

Jerzy Weyman (139), *Department of Mathematics, Northeastern University, Boston, Massachusetts 02115*

Contributors

Introduction

Free resolutions arise from systems of linear equations over rings other than fields. A system of linear equations in finitely many unknowns over a field has a basic set of solutions in terms of which all others can be expressed as linear combinations, and these basic solutions can be chosen to be linearly independent. Over a polynomial ring, or more generally any Noetherian ring R, a system of linear equations in finitely many unknowns still has a finite system of solutions in terms of which all others may be expressed, but now these solutions cannot in general be taken to be linearly independent. To find the dependence relations on a given system of solutions requires solving a new system of linear equations. Iterating this process, one gets a whole series of systems of equations, which makes up a **free resolution** of the original problem. If one considers the cokernel M of the matrix expressing the original system of equations as a module over R, one speaks of a free resolution of M. The free resolution expresses certain properties which are implicit in, but not at all obvious from, the original system of equations.

If the ring R is a polynomial ring or power series ring, then free resolutions can be taken to be of finite length (that is, the systems of equations produced eventually have linearly independent sets of basic solutions); this is essentially the celebrated "Hilbert Syzygy Theorem", which started the subject off. But over more general rings, the free resolution may involve infinitely many sets of equations.

Free resolutions and questions related to them occur in many areas of commutative algebra and algebraic geometry. In May of 1990 there was a small and informal conference in Sundance, Utah, organized by David Eisenbud, Craig Huneke, and Robert Speiser, on the topic of free resolutions and their uses in commutative algebra and algebraic geometry.

A good deal of the conference was devoted to discussions of the current state of work on some of the central problems in the area. These discussions seemed worth transmitting to a broader audience, and we were able to convince a number of the participants to write up accounts of areas in which they are expert. Some of these writeups develop groups of current problems which seem likely to influence future development of the field. Others are basic expositions of areas of current interest; and some contain new research, not otherwise published. A notion of the diversity and richness of the subject (and of the conference) can be obtained from a description of the topics treated:

The first group of papers treats basic questions about the relations between the systems of linear equations in a resolution. Much current interest relates to the size and ranks of these intermediate systems of equations (the "ranks of syzygies"), a subject related to questions about possible ranks of vector bundles on projective spaces. A central test case is that of a resolution of a module of finite length over a polynomial ring, and the state of our current knowledge and conjectures about this case are surveyed by Charalambous and Evans. Avramov treats the case where the free resolution is infinite, concentrating on what regularity may eventually appear in the resolution. Miller surveys the state of our knowledge about multiplicative structures on resolutions. In the most basic case, if the module M is of the form R/I, then the multiplicative structure of R/I extends (nonuniquely) to a commutative but only homotopy-associative algebra structure on the resolution. In some cases this multiplicative structure allows one to compare different parts of the resolution. The results one gets are closely related to the (originally geometric) notion of **linkage**, which plays an important role. Kempf briefly surveys some cases that correspond to the best possible behavior of an infinite free resolution of a graded module over a graded ring. These are the cases in which the systems of linear equations only involve linear polynomials. Surprisingly many resolutions of geometric interest have this property; noone yet really understands why.

The next pair of papers concerns Green's conjecture and some variants of it. Green's conjecture connects rather subtle geometric properties of an algebraic curve with the shape of the free resolution of the homogeneous coordinate ring of the curve. Eisenbud presents a new algebraic version of the conjecture, and explains, from an algebraic point of view, some of the motivation. He then surveys some of the approaches that have been tried and the partial results that they have yielded. In particular, there are several approaches which lead to absolutely explicit (but large) numerical matrices; determination of the ranks of these matrices would settle a leading case of the conjecture. Although several people had worked out such approaches, till now none of them has been written up. In the second paper, Bayer and Stillman exhibit two such matrices, from their own research, and explain the derivation of them. All those who are good at linear algebra, take note! Here is an opportunity to solve a central problem in the theory of Riemann surfaces by elementary methods.

The last five papers concern another range of algebraic and geometric topics:

Craig Huneke explains what is known and conjectured about the mysterious and interesting local cohomology modules $H_I^j(M) := \varinjlim_n \operatorname{Ext}_R^j(R/I^n, M)$ with respect to an arbitrary ideal I.

Cremona transformations are by definition the birational automorphisms of projective spaces. Sheldon Katz explains how free resolutions can be used for constructing and testing examples in this very classical geometric subject.

Paul Roberts gives a survey of work in the area of the "homological conjectures", where much has been done since Hochster's famous CBMS notes of 1975.

Bernd Ulrich treats "residual intersections": The central problem here is to say something about some of the (primary) components of an ideal I, given information about the ideal itself and information about the other components. This problem generalizes the problem of linkage, which is essentially the case in which I is a complete intersection.

Finally, Jerzy Weyman explains the current problems on invariants connected with the free resolutions of ideals associated to strata in representations, and some of the ideas of Kempf in this direction.

The conference was made possible by the generous support of Brigham Young University, in Provo, Utah. We personally found the conference highly productive, and the place and atmosphere unusually pleasant and congenial. Robert Speiser, our coorganizer, took charge of all the local arrangements with great spirit and success; the conference ran smoothly through the lubrication of his hard work. We, and the other participants, are grateful to the University for underwriting the conference, and also to the NSF for providing travel funds for some of the participants.

David Eisenbud and Craig Huneke   August 1991

Dept. of Mathematics Brandeis University Waltham MA 02254

eisenbud@brandeis.bitnet

Dept. of Mathematics Purdue University W. Lafayette IN 47907

clh@macaulay.math.purdue.edu

Structure and Size
of Free Resolution

Problems on infinite free resolutions

LUCHEZAR L. AVRAMOV [1,2]

Institute of Mathematics, ul. "Akad. G. Boncev" blok 8, 1113 Sofia, Bulgaria

Throughout this survey R will denote either a noetherian local ring with maximal ideal $\mathbf{m}$ and residue field k, or a graded algebra $R = \oplus_{n \geq 0} R_n$ with (irrelevant) maximal ideal $\mathbf{m} = \oplus_{n \geq 1} R_n$ and $R_0 = k$. The minimal number of generators of $\mathbf{m}$ is called the *embedding dimension* of R and denoted $\operatorname{edim} R$. A local (resp., graded) ring Q with maximal ideal $\mathbf{n}$ is said to be a *deformation of* R *of codimension* c if $R \cong Q/(x_1, \dots, x_c)$ for a Q-regular sequence $\mathbf{x} = x_1, \dots, x_c$, which is assumed to be homogeneous when R is a graded ring; in case $\mathbf{x}$ is contained in $\mathbf{n}^2$, the deformation is said to be *embedded*. If R is a local ring whose $\mathbf{m}$-adic completion $\hat{R}$ can be deformed to a regular local ring Q, then R is said to be a *complete intersection*; in the graded case, this notion refers to a graded ring R which can de deformed to a graded polynomial ring Q.

Only finitely generated R-modules will be considered; in the graded case homomorphisms of modules will be assumed to be compatible with the gradings and homogeneous of degree zero. A free resolution

$$(\mathbb{F}, \partial) : \ \dots \longrightarrow F_n \xrightarrow{\partial_n} F_{n-1} \longrightarrow \ \dots \ \longrightarrow F_1 \xrightarrow{\partial_1} F_0 \ (\longrightarrow M \longrightarrow 0)$$

is said to be minimal if $\partial_n(F_n) \subset \mathbf{m}F_{n-1}$ for $n \geq 1$. It is well known that any two minimal resolutions of M are isomorphic as complexes of R-modules. Thus, the module $\partial_n(F_n)$ is defined uniquely up to isomorphism: it is called the n'th *syzygy* of M and denoted $\operatorname{Syz}_n^R(k)$. Similarly, the rank of the free R-module F_n does not depend on the choice of $\mathbb{F}$: it is called the n'th *Betti number* of R and denoted $b_n^R(M)$. Betti numbers are usually computed (and frequently defined) via the equalities $b_n^R(M) =$

---

[1] Participation at the Sundance Conference on Free Resolutions in Algebraic Geometry and Commutative Algebra was supported by the NSF under the U.S.-Bulgarian Program in Algebra and Geometry.

[2] Present address: Department of Mathematics, Purdue University, West Lafayette, IN 47907.

$\dim_k \operatorname{Tor}_n^R(M, k) = \dim_k \operatorname{Ext}_R^n(M, k)$. It is often advantageous to study Betti numbers wholesale rather than individually, and the standard vehicle for the relevant information is the formal power series $P_M^R(t) = \sum_{n\geq 0} b_n^R(M)t^n$, called the *Poincaré series* of M ; by traditional abuse of language the series $P_k^R(t)$ is also called the Poincaré series of R .

1. Poincaré series

The conjecture that $P_k^R(t)$ always represents a rational function in t has been a powerful motivation for the study of infinite free resolutions for almost 25 years. However, in 1979 Anick [1] constructed an artinian (graded) ring with $\mathbf{m}^3 = 0$, embedding dimension 5 , and length 13 , such that $P_k^R(t)$ is transcendental. A related construction was then proposed by Löfwall and Roos [41]. Building up on this, Bøgvad [21] produced an artinian (graded) Gorenstein ring with transcendental Poincaré series: it has $\mathbf{m}^4 = 0$, embedding dimension 12 , and length 26. Instructive surveys on the (non-)rationality of $P_k^R(t)$ have been given by Roos [47] and by Babenko [16].

The negative results notwithstanding, quite general statements on rationality are known for several classes of local rings. Namely, the formal power series $P_M^R(t)$ is rational for every finitely generated R-module M when R satisfies one of the following conditions:

(a) R is a complete intersection [31];

(b) R is a Golod ring [29];

(c) edim R – depth R $\leqslant$ 3 [15];

(*)

(d) edim R – depth R = 4 , and R is Gorenstein [36];

(e) R is directly linked to a complete intersection [15];

(f) R is linked in two steps to a complete intersection, and R is Gorenstein [15].

Our first problem (which can be refined when R is Gorenstein) puts the focus on the gaps which still remain between the cases when rationality or irrationality is known :

Problem 1. Find "natural" sufficient conditions on R for $P_M^R(t)$ to be a rational function for each R-module M . In particular test the following properties: edim R - depth R = 4 ; length R $\leqslant$ 12 ; R is in the linkage class of a complete intersection. As a first step, do they imply the rationality of the Poincaré series of R ?

Partial results are available: Backelin and Fröberg [18] have obtained rational expressions for $P_k^R(t)$ when R is graded of length $\leqslant$ 7, and Palmer [42] has proved that $P_M^R(t)$ is rational for all M over certain almost complete intersections with edim R - depth R = 4 , cf. also the note at the end of the paper. However, Jacobsson [35] has constructed a ring R with rational $P_k^R(t)$, but which has a module M with $P_M^R(t)$ transcendental. If $P_M^R(t)$ is rational for all M , then the next problem asks whether there can be an infinite number of essentially different Poincaré series.

Problem 2. Assume that R has the property that the Poincaré series of any finitely generated R-module is rational. Does a polynomial $Den^R(t)$ with integer coefficients then exist, such that for each R-module M the series $Den^R(t)P_M^R(t)$ is a polynomial ?

In all cases when the hypothesis of Problem 2 is known to hold, the answer is positive, and furthermore one can take $Den^R(t)$ to be the denominator of $P_k^R(t)$ written as an irreducible rational function. This extra information has been used in an essential way in [9] for the study of a conjecture of Eisenbud on modules with bounded Betti numbers, cf. Section 2 below.

It is very interesting to study quantitatively the gap between rational and non-rational series. At least for equicharacteristic rings, this can be put in concrete terms as follows. Let $R = k \oplus m$ be a finite dimensional k-algebra with nilpotent ideal m , and let V denote an n-dimensional k-vector space. A bilinear pairing $\mu: R \times V \to V$ will be identified with an element of $R^* \otimes V^* \otimes V$. Those μ which define on V a structure of (left) R-module form an algebraic subset $\mathbf{Mod}_n^R(k)$ of $R^* \otimes V^* \otimes V$. The linear group $GL(n, k)$ acts in an obvious way on $\mathbf{Mod}_n^R(k)$ and the orbits of this action are in one-to-one correspondence with the isomorphism classes of R-modules M of dimension n over k. There is a single closed orbit, given by the module k^n (for details cf. e.g. [27]).

Problem 3. Study the geography of the set $\{M \in \mathbf{Mod}_n^R(k) \mid P_M^R(t) \text{ is rational}\}$: When is it non-empty? If it is, then does it (or its complement) contain a non-empty open subset? Does there exist an irreducible component consisting of modules with rational (or non-rational) series? What is the maximal dimension of an orbit which contains such a module?

There are versions of these questions for the set $\mathbf{LocAlg}_{n,m}(k) \subset V^* \otimes V^* \otimes V$ which parametrizes the commutative k-algebra structures on $k \oplus V$ such that $V^m = 0$, $2 \leqslant m \leqslant n+1$ (cf. e.g. [34] for a description of the corresponding constructions, and the recent article [48] for information on the components of $\mathbf{LocAlg}_{n,3}(k)$). For example, what can be said about $\mathbf{LocAlg}_{12,3}(k)$? (By Anick's example this is the first case where it is known that a non-trivial situation occurs.) Note that in these questions the restriction to artinian rings and finite length modules is natural in view of results of Levin [40] which reduce rationality problems for Poincaré series to the artinian case.

2. Bounded Betti numbers

The projective dimension of a non-zero R-module M may be introduced by the formula: $\text{pd}_R M = \sup \{n \in \mathbb{N} \mid b_n^R(M) \neq 0\}$. From this point of view, the simplest modules of infinite projective dimension are those whose Betti numbers are bounded. They have received some attention in literature, but very natural and important questions are still open, even the existential one:

Problem 4. Which rings have modules of infinite projective dimension with bounded Betti numbers?

That the answer has to be non-trivial is shown for instance by the existence of rings R such that the **Betti** numbers **s**trictly **i**ncrease for each non-free R-module M and for $n \geqslant 1$ (in [44] such R are introduced by the name BNSI rings). For an elementary example consider a ring R with $m \neq m^2 = 0$. The first syzygy of a non-free module M is then a direct sum of copies of $\mathbf{k}$. As it is easily seen that $b_n^R(\mathbf{k}) =$ (edim R)^n for $n \geqslant 0$, it follows that R is a BNSI ring when edim $R \geqslant 2$. More general constructions of BNSI rings are given by Ramras [44], Gover and Ramras [30], and Lescot [37].

As a potential source of bounded Betti sequences, consider periodic modules: if m is a positive integer, then a non-zero R-module M is said to be *periodic of period* m if it is isomorphic to its m'th syzygy (and then $Syz_n^R(M) \cong Syz_{n+m}^R(M)$ for $n \in \mathbb{N}$). The Betti numbers of a periodic module are obviously bounded, and the converse is true when R is artinian with $\mathbf{k}$ is algebraic over a finite field, cf. [28]. Thus, a step in answering the question raised in the preceding problem will be a solution of the following one.

Problem 5. Characterize the rings which have a periodic module.

The only general construction for both problems seems to be for rings which have an embedded (codimension one) deformation $R \cong Q/(x)$ (cf. [26] for regular Q, and [33] or [13] otherwise). There are then homomorphisms of free Q-modules $\varphi: G \longrightarrow F$ and $\psi: F \longrightarrow G$ such that: $rank_R F = rank_R G$; $\varphi(G) \subset \mathbf{n}F$ and $\psi(F) \subset \mathbf{n}G$; $\varphi\psi = x\mathrm{id}_F$ and $\psi\varphi = x\mathrm{id}_G$; the sequence of free R-modules:

$$(**) \quad \ldots \xrightarrow{\psi \otimes_Q R} G \otimes_Q R \xrightarrow{\varphi \otimes_Q R} F \otimes_Q R \xrightarrow{\psi \otimes_Q R} G \otimes_Q R \xrightarrow{\varphi \otimes_Q R} F \otimes_Q R \longrightarrow \ldots$$

is exact. In Eisenbud's terminology M = Coker φ is obtained from a *matrix factorization* of x ; such a module is periodic of period 2 and has constant Betti numbers.

If R is complete with infinite residue field, then a series of theorems establishes that under any of the conditions (∗) the (depth R + 1)'st syzygy of a module with bounded Betti numbers is obtained from some matrix factorization: for complete intersections cf. [26]; for Golod rings which are not complete intersections, it is known that the sequence { $b_n^R(M)$ | n > edim R } is strictly increasing, cf. [37]; the remaining cases are covered by [9]: cf. the discussion following Problem 7 below for more details.

However, periodic modules also exist over some rings which admit no nontrivial deformation, cf. [12]. Furthermore, it is known that the solutions to Problems 4 and 5 do not coincide in general: Gasharov and Peeva [28] have exhibited artinian rings of embedding dimension 4 and artinian Gorenstein rings of embedding dimension 5 which have modules with constant Betti numbers, but with periods different from 2 , or with no periodic pattern in their resolutions (in view of the quoted results these examples are minimal with respect to edim R - depth R) . Thus it is hard at present to conjecture a reasonable general answer to either of the last two problems. The first concrete question to consider is whether over a ring R in the linkage class of a complete intersection each module with bounded Betti numbers has a syzygy obtained from a matrix factorization.

Before leaving periodicity, note a very down-to-earth reformulation of Problem 5. Indeed, if M is periodic of period m , then $N = \oplus_{i=0}^{m-1} Syz_n^R(M)$ is periodic of period 1, that is, it is isomorphic to each one of its syzygies. Thus, the last problem can be restated in entirely non-homological terms as follows:

Problem 5′. Give necessary and sufficient conditions on R for the existence of a non-zero square matrix A with coefficients in **m** , such that Im A = Ker A .

The next question — first raised by Ramras [45] — is about the nature of bounded sequences of Betti numbers:

Problem 6. Does the Betti sequence of a module M with bounded Betti numbers become eventually constant?

Note that if the Poincaré series of M is rational, then M has bounded Betti numbers if and only if $P_M^R(t)$ can be written in the form $p(t)/(1-t^m)$ for some integer polynomial $p(t)$ and some positive integer m, cf. e.g. [9]. It follows, in particular, that the Betti sequence of M is eventually periodic. (The converse is obvious: the Poincaré series of a module with eventually periodic Betti numbers is rational of the form described above.) Thus, before tackling the preceding question, one might try to answer the following one:

Problem 7. Does a bounded Betti sequence eventually become periodic?

A positive answer to Problem 6 (and hence to Problem 7 as well) is available for the modules over the rings described in (*). The proof, obtained in [9], proceeds to show first that over these rings any module with bounded Betti numbers has finite *virtual projective dimension*, that is, the number $\text{vpd}_R M = \inf \{\text{pd}_Q \hat{M} \mid Q$ ranges over all deformations of $\hat{R}\}$ is finite (there are some further technicalities when the residue field of R is finite). The conclusion is then obtained by applying a result of [8], which shows that if $\text{vpd}_R M < \infty$ and M has bounded Betti numbers, then the (depth R – depth$_R$ M + 1)'st syzygy of M is obtained from a matrix factorization.

There is yet another aspect to the existence of periodic modules. Indeed, a periodic resolution can be extended not only backwards but also forwards, thus exhibiting a periodic module M as an n'th syzygy for any $n \in \mathbb{N}$. A non-zero module which has this property will be called an *infinite syzygy*.

Problem 8. Characterize the rings which have an infinite syzygy module.

A version of this problem is considered in [11]. The following remarks summarize what seems to be known on the subject. Clearly, infinite syzygies cannot exist over regular local rings, or over BNSI rings. On the other hand, it is seen from (**) that infinite syzygies exist over any ring which has a non-trivial embedded deformation. Note also that an infinite syzygy module M is a (depth R + 1)'st syzygy of a finitely generated non-zero R-module, hence the Auslander-Buchsbaum equality shows that the projective dimension of M may not be finite.

An infinite syzygy module M over a singular Cohen-Macaulay ring is maximal Cohen-Macaulay, that is, $\text{depth}_R M = \dim R$. Conversely, over a Gorenstein ring R any non-free maximal Cohen-Macaulay module is an infinite syzygy. Indeed, the most natural way to construct an infinite syzygy (over an arbitrary ring R) is to find a reflexive module M such that $M^* = \text{Hom}_R(M,R)$ has infinite projective dimension and satisfies $\text{Ext}_R^n(M^*, R) = 0$ for n > 0 : the R-dual of a minimal resolution of M^* then shows $M \cong M^{**}$ is an infinite syzygy. Recall now that in Auslander and Bridger's [3] theory of G-dimension, a module M is said to be of G-dimension zero if it is reflexive and satisfies $\text{Ext}_R^n(M^*, R) = 0 = \text{Ext}_R^n(M, R)$ for n > 0 . Thus, if $\text{G-dim}_R M = 0$, then M is an infinite syzygy provided $\text{pd}_R M^* = \infty$. Furthermore, it is proved in [3] that for any module N over a Gorenstein ring R there is an equality $\text{G-dim}_R N + \text{depth}_R N = \dim R$. It follows that if M is a maximal Cohen-Macaulay module, then $\text{G-dim}_R M = 0$. As M^* also is then maximal Cohen-Macaulay, $\text{pd}_R M^*$ is finite only if M^* is free, and then $M \cong M^{**}$ is free. Thus a non-free maximal Cohen-Macaulay R-module M is reflexive and has $\text{pd}_R M^* = \infty$, which establishes our assertion.

Since infinite syzygies have been constructed over some artinian (graded) rings which are not Gorenstein and have no deformations, cf. [12], Problem 8 is open even for artinian rings.

3. Asymptotic behavior of Betti sequences

If M is an R-module of infinite projective dimension, then for its (depth R)'th syzygy N there exists a maximal R-regular sequence **y** which is also N-regular. It follows that for $n \in \mathbb{N}$ there are equalities $b_{n+\text{depth}R}^R (M) = b_n^R(N) = b_n^{R/(\mathbf{y})}(N/(\mathbf{y})N)$. Thus, when dealing with asymptotic properties of Betti sequences one may assume that the ring R has depth zero. The first question on unbounded Betti sequences seems to be whether they also satisfy $\lim_{n\to\infty} b_n^R(M) = \infty$, cf. [45]. The following one, proposed in [6], asks for more and includes Problem 6:

Problem 9. Is the Betti sequence of an arbitrary R-module eventually non-decreasing?

One might try to start with modules of finite length, where length considerations have often proven effective, cf. [46, 28, 25]. Before describing cases where the answer to Problem 9 is known, we shortly discuss the asymptotic behavior of sequence of Betti numbers.

The simplest estimate on such a sequence is given as follows: M is said to have *complexity* d , denoted $cx_R M = d$, if d is the least integer such that $b_n^R(M) \leqslant \alpha n^{d-1}$ holds for some positive $\alpha \in \mathbb{R}$ and for n >> 0 , cf. [8]. A module M is said to have *strong polynomial growth of degree* d if there are polynomials p(X) and $q(X) \in \mathbb{R}[X]$ which are both of degree d , have the same leading term, and satisfy $p(n) \leqslant b_n^R(M) \leqslant q(n)$ for n >> 0 . Similarly, M is said to have *strong exponential growth* if there is a real numbers $\alpha > 1$ such that $\alpha^n \leqslant b_n^R(M)$ for n >> 0 . We refer to [9] for a discussion of related conditions, as well as for a proof of the well-known fact that for any module M there is a $\beta \in \mathbb{R}$ such that $b_n^R(M) \leqslant \beta^n$ for n >> 0 . The following questions are taken from [9]; note that in the first unknown case - that of complexity 1 - the last question coincides with that raised in Problem 6.

Problem 10. Does every R-module have one of the two types of growth described above? If the complexity of M is finite, does the inequality $cx_R M \leqslant edim R - depth R$ then hold? Does a module of finite complexity have strong polynomial growth?

Answers to both problems are known in roughly the same situations: when $M = k$ [49, 32, 7]; when R is artinian of length $\leqslant 7$ or artinian Gorenstein of length $\leqslant 11$ [28]; when $\mathbf{m}^3 = 0$ [37]; when M has finite virtual projective dimension, in particular when R is a complete intersection [8, 13]; when R is a Golod ring [39, 9]. Although

some of the proofs exploit the rationality of $P_M^R(t)$, it seems that the growth and rationality properties of Betti sequences are in general unrelated: witness the result of Anick
and Gulliksen [2], which for any ring R produces a ring R' with $(\mathbf{m}')^3 = 0$, such that
the Poincaré series of either one is rationally expressible in terms of the other one's.
Growth problems in some cases in which rationality is not known have been treated by
Choi [24, 25].

As a ring with edim R – depth R $\leqslant$ 2 is either a complete intersection or a Golod
ring, Problems 9 and 10 then have positive solutions. The same conclusion holds for
Gorenstein rings with edim R – depth R = 3, since then either R is a complete intersection or there is an odd integer r $\geqslant$ 5 such that every R-module has a rational
Poincaré series with denominator $1 - rt^2 - rt^3 + t^5$. These observations however do not
suffice to prove the growth of Betti numbers for all the rings described by conditions (*)
in Section 1. As for the asymptotics of Betti sequences, in this case they are "almost"
known, in the sense that the asymptotics of the first sum-transform $\beta_n^R(M) = \sum_{i=0}^n b_i^R(M)$ are determined in [9].

The scarcity of results and the considerable effort spent in proving those described
above illustrate both the apparent difficulty of the problems and the extent of our igno-
rance concerning infinite resolutions. While the rings described in (*) might yield to already available techniques, there seems to be a need for a whole new arsenal of tools to
tackle such questions in general: we shall come back to this in the next section.

Another useful measure of the asymptotic growth of the Betti sequence of M is
given by the radius of convergence of its Poincaré series, that is, the real number $\rho_R(M)$
= $1/\limsup (b_n^R(M))^{1/n}$. We call it the *radius of convergence of* M ; note that $\rho_R(M)$ =
∞ if and only if $pd_R M < \infty$, and that otherwise there are inequalities $0 < \rho_R(M) \leqslant 1$.

In general, the modules whose radius of convergence is extremal seem particularly
interesting. It is known that the inequality $\rho_R(M) \geqslant \rho_R(k)$ holds for all M. This follows from the more precise result, that for each M there is a positive real number γ
such that the inequality $b_n^R(M) \leqslant \gamma b_n^R(k)$ holds for all $n \in \mathbb{N}$. When M is artinian,

the assertion is immediate from the long homology exact sequence, which shows that one can then take γ equal to the length of M. The general case is reduced to the artinian one by a result of [4] and [40] which proves that for each M there is a positive integer r such that the equality $b_n^R(M/\mathbf{m}^r M) = b_n^R(M) + b_{n-1}^R(\mathbf{m}^r M)$ holds for all n. At the other extreme, it is clear that a module of finite non-zero complexity has radius of convergence 1, and it has been proved that $\rho_R(\mathcal{k}) \geqslant 1$ if and only if R is a complete intersection, cf. [6, 7].

Finally, note that the radius of convergence of a module is equal to that of any of its syzygies, and that for each positive real number ρ the R-modules M which satisfy the inequality $\rho_R(M) \leqslant \rho$ form, in the category of all finitely generated R-modules, a subcategory which is closed under extensions. Thus the radius of convergence may be useful in studying the Grothendieck group G(R) of a local ring R.

The preceding remarks provide the context for the next group of questions.

Problem 11. Does $\rho_R(M) \geqslant 1$ imply that M has finite complexity? Is there a constant $\rho_R < 1$, such that $\rho_R(M) < 1$ implies $\rho_R(M) < \rho_R$? Is there a ring for which the set $\{\rho_R(M)\}$, where M ranges over all finitely generated R-modules, is infinite?

We finish this section by taking a quick look at the initial portions of Betti sequences. While for modules of finite projective dimension the whole problem is here, in case of infinite dimension this piece of the resolution seems to be almost beyond control. This adds interest to Huneke's remark (at the Sundance Conference), that the following question, which for modules of finite projective dimension has recently focused much attention and is discussed elsewhere in these Proceedings, cf. [23], is also open in general:

Problem 12. Does the inequality $b_n^R(M) \geq \binom{\dim R - \dim M}{n}$ hold for all n?

4. Structures on resolutions

By general nonsense, Yoneda pairings provide $\mathrm{Ext}_n^*(\mathit{k}, \mathit{k})$ with a structure of graded k-algebra, and $\mathrm{Ext}_R^*(M, \mathit{k})$ with a structure of a graded left module over it. The key to some of the results described in the preceding sections has been provided by an understanding these pairings at a non-trivial level. The information on the Yoneda algebra is concentrated in a much smaller object functorially attached to R , the *homotopy Lie algebra* $\pi*(R)$. This is a graded Lie (super-)algebra whose universal enveloping algebra is $\mathrm{Ext}_R^*(\mathit{k}, \mathit{k})$, cf. [6]. Thus, the cohomology module of M is in a natural way a graded representation of the homotopy Lie algebra. Very simple structural questions are still unanswered, for example the following one:

Problem 13. For what M is the $\mathrm{Ext}_R^*(\mathit{k}, \mathit{k})$-module $\mathrm{Ext}_R^*(M, \mathit{k})$ finitely generated?

Even when the cohomology module is known to be finitely generated, it might still be difficult to gather from this fact numerical information, due to the complexity of the homotopy Lie algebra $\pi*(R)$, which is conjectured to contain a non-abelian free subalgebra unless R is a complete intersection, cf. [5]. An original and sometimes efficient way to circumvent this inconvenient has been proposed by Backelin and Roos [20], who investigate the cohomology of $\mathrm{Ext}_R^*(M, \mathit{k})$ over the cohomology algebra of $\mathrm{Ext}_R^*(\mathit{k}, \mathit{k})$, which is a (skew-)commutative graded k-algebra. On the other hand, there are indications of correlations between finite complexity and restrictions on the structure of the cohomology module of M . For instance, the theory of modules of finite virtual projective dimension [8] is entirely based on the portrait of M provided by its cohomology module, which in this case is finitely generated over a central polynomial subalgebra of $\mathrm{Ext}_R^*(\mathit{k}, \mathit{k})$. Thus, in some cases the action of a large part of the Yoneda algebra may be irrelevant. The next problem takes this observation into account.

Problem 14. When is $\text{Ext}_R^*(M, k)$ finitely generated over a *noetherian* subalgebra or factor algebra of $\text{Ext}_R^*(k, k)$?

Note that Bøgvad and Halperin [22] have proved that if $\text{Ext}_R^*(k, k)$ itself is noetherian, then R is a complete intersection (this had been conjectured by Roos; the converse is also true and had been known for some time).

If M is periodic of period m , then the initial portion of its minimal resolution provides a length m exact sequence:

$$0 \longrightarrow M \longrightarrow F_{m-1} \xrightarrow{\partial_{m-1}} F_{m-2} \quad \cdots \cdots \longrightarrow F_1 \xrightarrow{\partial_1} F_0 \longrightarrow M \longrightarrow 0 .$$

Right multiplication by the congruence class $\mu \in \text{Ext}_R^m(M, M)$ of this exact sequence provides isomorphisms $\text{Ext}_R^n(M, N) \rightarrow \text{Ext}_R^{n+m}(M, N)$ for each R-module N and for $n \in \mathbb{N}$. If one tries to find an explanation of periodicity phenomena in the structure of $\text{Ext}_R^*(M, k)$, the next problem arises (its solution is positive for modules of finite virtual projective dimension, hence for all modules over the rings described by the conditions (*) in Section 1).

Problem 15. If M is periodic of period m , does there exist $\chi \in \text{Ext}_R^m(k, k)$ such that left multiplication by χ provides isomorphisms $\text{Ext}_R^n(M, k) \rightarrow \text{Ext}_R^{n+m}(M, k)$ for $n \in \mathbb{N}$?

The $\text{Ext}_R^*(k, k)$-module structure on $\text{Ext}_R^*(M, k)$ reflects the "linear part" of the differential of $(\mathbb{F}, \partial)$: for $M = k$ an explanation of this admittedly cryptic statement may be found in [6]. The examination of the Yoneda Ext-algebra from the point of view of its homotopy Lie algebra nucleus is strongly connected with ideas and techniques from rational homotopy theory, which are discussed at length in [6] and [14]. This approach also suggests that the differential of a minimal free resolution might be "reconstructed" from some higher order structures.

One such structure was associated by Eisenbud [26] with a deformation of R . When $R = Q/(x)$ for a non-zero divisor x , let $(\tilde{\mathbb{F}}, \tilde{\partial})$ be a sequence of homomorphisms of free Q-modules such that $(\widetilde{F} \otimes R, \widetilde{\partial} \otimes R) = (F, \partial)$. Since $(\widetilde{\partial} \otimes R)^2 = \partial^2 = 0$, there is a unique degree -2 endomorphism $\widetilde{t}$ of the *graded* Q-*module* $\widetilde{F}$, such that $\widetilde{\partial}^2 = x \widetilde{t}$. Using the fact that x is Q-regular it is easy to see that $t = \widetilde{t} \otimes R$ is an endomorphism of the *complex* (F, ∂). In [26] a more general version of this construction is presented: it associates to a codimension c deformation a family of endomorphisms $t_1, \dots, t_c$ of (F, ∂), which are unique up to homotopy and commute up to homotopy. We call these endomorphisms the *Eisenbud operators* on (F, ∂).

Problem 16. Does the minimal resolution of M admit a family of *commuting* Eisenbud operators, that is, does (F, ∂) support a structure of differential graded module over the ring $R[t_1, \dots, t_c]$ graded by assigning to the t_i's degree -2 ?

The interest of a positive answer, conjectured by Eisenbud in [26], is that when $pd_Q M < \infty$ this module structure is cofinite in the sense that the graded $R[t_1, \dots, t_c]$-module $F^* = Hom_R(F, R)$ is finitely generated. Thus, even though the minimal R-free resolution of M is infinite, for each R-module N the graded R-module $Ext_R^*(M, N) = H(Hom_R(F, N)) \cong H(F^* \otimes_R N)$ is determined by a finite set of data: the action of the differential on an initial portion of the resolution and a structure of finitely generated module over a noetherian graded polynomial ring. It was recently proved in [13] that whenever $c \geqslant 2$ there are R-modules for whose resolution there exist no commuting set of Eisenbud operators. However, the obstruction constructed in that paper vanishes for all sufficiently high syzygies of any module, hence Eisenbud's conjecture is still open in its important asymptotic aspect.

5. Graded rings

The treatments of the local and graded cases have so far been completely parallel. Next we focus on problems specific to the graded case, where more structure is available; in order to concentrate on the most important case, for the rest of this section R denotes a graded ring generated by its elements of degree 1. If M is a graded R-module, set $i(M) = \inf\{i \mid M_i \neq 0\}$. As usual, the q'th twist $M[q]$ of M is defined for $q \in \mathbb{Z}$ by the formula $(M[q])_i = M_{i+q}$, $i \in \mathbb{Z}$. The formal Laurent series $\mathrm{Hilb}_M(u) = \sum_{n \in \mathbb{Z}} \dim_k M_n u^n \in \mathbb{Z}((u)) = \mathbb{Z}[[u]][u^{-1}]$ is called the *Hilbert series of* M: it is of course rational, with denominator $(1-t)^{\dim M}$ for its irreducible form.

In the graded case there is a second grading on the resolution $(\mathbb{F}, \partial)$, coming from the fact that each free graded R-module F_n is isomorphic to $\oplus_{s \in \mathbb{Z}} R[-q_{ns}]$. This induces for each n a structure of graded k-vector space on $\mathrm{Tor}_n^R(M, k) = \oplus_{s \in \mathbb{Z}} \mathrm{Tor}_n^R(M, k)_s$. The integers $b_{ns}(M) = \dim_k \mathrm{Tor}_n^R(M, k)_s$ are called the *graded Betti numbers of* M. It is well known and easily seen that $b_{n,t+i(M)}(M) = 0$ for $t < n$ and that $b_{ns}^R(M) = 0$ for $s \gg 0$, hence for $n \in \mathbb{N}$ the sum $\sum_{s \in \mathbb{Z}} b_{ns}^R(M)$ is finite and equals $b_n^R(M)$. Similarly, $P_M^R(t,u) = \sum_{n \in \mathbb{Z}, s \in \mathbb{Z}} b_{ns}^R(M) t^n u^s \in \mathbb{Z}[[t]]((u))$ is a formal power series in the variable t and a formal Laurent series in the variable u, called the *double Poincaré series of* M *over* R. From it one recovers the usual Poincaré series via the equality $P_M^R(t) = P_M^R(t,1)$, but also the Hilbert series via the equality $\mathrm{Hilb}_M(u) = P_M^R(-1,u) \mathrm{Hilb}_R(u)$.

Since the examples of irrational $P_k^R(t)$ mentioned in Section 1 are for graded artinian rings, $P_M^R(t)$ is not in general a rational function, and then neither is $P_M^R(t,u)$. However, in the extremal situation when R is *multigraded* (that is, R is a homomorphic image of a polynomial ring by an ideal generated by monomials in the indeterminates) and M is a multigraded R-module, then $P_M^R(t,u)$ is a rational function of two variables, as shown by Backelin [17] for $M = k$ and extended by Lescot [38] to finitely generated modules.

Problem 17. Does there exist an R-module M such that $P_M^R(t)$ is a rational function, but $P_M^R(t,u)$ is not?

The graded modules with the simplest possible resolutions are those which have $b_n(M) = b_{n,n+i(M)}(M)$ for $n \in \mathbb{N}$. This condition is equivalent to the requirement that

$P_M^R(t,u)/u^{i(M)}$ is a formal power series in the single variable tu, and it is also equivalent to the property that all the differentials in a minimal resolution of M are given by matrices of linear forms. The last characterization motivates the name *module with linear resolution* given to such M. Note that a module M with linear resolution has rational Poincaré series of a specific form, as shown by the equalities:

$$P_M^R(t) = P_M^R(t,1) = (-1)^{i(M)} P_M^R(-t,-1) = \pm \text{Hilb}_M(-t)/\text{Hilb}_R(-t) ,$$

and this implies that even the double Poincaré series of M is then a rational function.

A concept related to that of of linear resolution is introduced in terms of the *regularity* of M, defined by $\text{reg}_R M = \inf \{ r \mid \text{Tor}_n^R(M, k)_{n+j} = 0 \text{ for } j > r \text{ and } n \in \mathbb{N} \}$. Thus, a nonzero module M has finite regularity if and only if its graded Betti numbers are concentrated in a strip parallel to the diagonal, and it has a linear resolution if and only if $\text{reg}_R M[i(M)] = 0$.

Problem 18. Does there exist a graded module M of finite regularity, for which the Poincaré series $P_M^R(t)$, or even the double Poincaré series $P_M^R(t,u)$, is not a rational function?

Note that a nonzero module of finite projective dimension has finite regularity. I do not know whether the existence of an R-module of infinite projective dimension and finite regularity, or even of one with an infinite linear resolution, places any restriction on the ring R. However, such modules seem hard to come by, hence the next problem.

Problem 19. Which graded rings R have modules of infinite projective dimension and of finite regularity (respectively, with linear resolution)?

As usual in homological games, conditions on the residue field k have an especially strong impact. A class of graded rings which has kept emerging from different contexts and has attracted significant attention during the last few years is formed by the so-called *Koszul algebras* (which also go by the names *homogeneous Koszul algebras*, or

Priddy algebras). They are defined by the requirement that the residue field k is a module with linear resolution, and admit a variety of possible characterizations, cf. [43, 19]. References to different constructions of Koszul algebras (many of them homogeneous coordinate rings of "interesting" projective varieties) are given e.g. in [10]. The main result of that paper proves that for a module M over a Koszul algebra R there is an inequality $\mathrm{reg}_R M \leqslant \mathrm{reg}_Q M$ with Q denoting the polynomial k-algebra $\mathrm{Sym}_k R_1$. In particular, any module over a Koszul algebra has finite regularity. A strong converse is conjectured in [10], in the form of a positive answer to the next question. Once more, such an answer is available for the rings described by the conditions (*) of Section 1.

Problem 20. Does $\mathrm{reg}_R k < \infty$ imply that R is a Koszul algebra?

Finally, we address the influence of homological properties of the graded modules over a graded ring R on properties of modules over the localization R_m at the irrelevant maximal ideal **m**. Clearly, if $(\mathbf{F}, \partial)$ is a minimal graded resolution of a graded R-module M, then $(\mathbf{F_m}, \partial_m)$ is a minimal R_m-free resolution of M_m; in particular there are equalities $b_n^R(M) = b_n^{R_m}(M_m)$ for $n \in \mathbf{N}$, hence in the study of R_m-modules induced from graded R-modules one can use the extra help coming from the grading. What about the other ones? A sample question is formulated below:

Problem 21. If a graded ring R has the property that all graded R-modules have rational Poincaré series (respectively, have eventually non-decreasing Betti numbers), does then the local ring R_m have the same property?

References

[1]   D.J.Anick: A counterexample to a conjecture of Serre, Ann. of Math. **115** (1982), 1-33.

[2]   D.J.Anick and T.H.Gulliksen: Rational dependencies among Hilbert and Poincaré series,J.Pure Appl. Algebra **38** (1985), 135-157.

[3]   M.Auslander and M.Bridger: *Stable Module Theory*, Memoirs Amer. Math. Soc. **94**, AMS, Providence, RI, 1969.

[4]   L.L.Avramov: Small homomorphisms of local rings, J. Algebra 50 (1978), 400-453.

[5]   L.L.Avramov: Free Lie subalgebras of the cohomology of local rings, Trans. Amer. Math. Soc. **270** (1982), 589-608.

[6]   L.L.Avramov: Local algebra and rational homotopy, Asterisque 113-114 (1984), 15-43.

[7]   L.L.Avramov: Local rings of finite simplicial dimension, Bull. Amer. Math. Soc. (New Series) **10** (1984), 289-291.

[8]   L.L.Avramov: Modules of finite virtual projective dimension, Invent. Math. 96 (1989), 71-101.

[9]   L.L.Avramov: Homological asymptotics of modules over local rings, in: *Commutative Algebra,* MSRI Publ. **15** (Springer, New York - Berlin, 1989), 33-62.

[10] L.L.Avramov and D.Eisenbud: Regularity of modules over a Koszul algebra, J. Algebra (to appear).

[11] L.L.Avramov and H.-B.Foxby: Homological dimensions of unbounded complexes, J. Pure Appl. Algebra **71** (1991), 129-156.

[12] L.L.Avramov, V.N.Gasharov, and I.V.Peeva: A periodic module of infinite virtual projective dimension, J. Pure Appl. Algebra **62** (1989), 1-5.

[13] L.L.Avramov, V.N.Gasharov, and I.V.Peeva: Modules of finite virtual projective dimension. II. (in preparation).

[14] L.L.Avramov and S.Halperin: Through the looking glass: A dictionary between rational homotopy theory and local algebra, in : Lecture Notes in Math. **1183** (Springer-Verlag, Berlin, 1986), 1-27.

[15] L.L.Avramov, A.Kustin, and M.Miller: Poincaré series of modules over local rings of small embedding codepth or small linking number, J. Algebra **118** (1988), 162-204.

[16] I.K.Babenko: Problems of growth and rationality in algebra and topology, Usp. Mat. Nauk **41** (1986), No. 2 , 95-142 [in Russian]; English translation: Russ. Math. Surv. **41** (1986), No 2 , 117-175.

[17] J.Backelin: Les anneaux locaux à relations monomiales ont des séries de Poincaré-Betti rationnelles, C. R. Acad. Sci. Paris **295** (1982), Ser. I, 605-610

[18] J.Backelin and R.Fröberg: Poincaré series of short artinian rings, J. Algebra 96 (1985), 495-498.

[19] J.Backelin and R.Fröberg: Koszul algebras, Veronese subrings and rings with linear resolutions, Rev. Roumaine Math. Pures Appl. 30 (1985), 85-97.

[20] J.Backelin and J.-E.Roos: When is the double Ext-algebra of a local noetherian ring again noetherian?, in: Lecture Notes in Math. 1183 (Springer-Verlag, Berlin, 1986), 101-119.

[21] R.Bøgvad: Gorenstein rings with transcendental Poincaré series, Math. Scand. 53 (1983), 5-15.

[22] R.Bøgvad and S.Halperin: On a conjecture of Roos, in: Lecture Notes in Math. 1183 (Springer-Verlag, Berlin, 1986), 120-127.

[23] H.Charalambous and E.G.Evans, Jr.: Problems on Betti numbers of finite length modules, these Proceedings.

[24] S.Choi: Betti numbers and the integral closure of ideals, Math. Scand. 66 (1990) 173-184.

[25] S.Choi: Exponential growth of Betti numbers, J. Algebra (to appear).

[26] D.Eisenbud: Homological algebra on a complete intersection, with an application to group representations. Trans. Amer. Math. Soc. 260 (1980), 35-64.

[27] P.Gabriel: Finite representation type is open, in: Lecture Notes Math. 488 (Springer-Verlag, Berlin, 1975), 132-155.

[28] V.N.Gasharov and I.V.Peeva: Boundedness versus periodicity over local rings, Trans. Amer. Math. Soc. 320 (1990), 569-580.

[29] F.Ghione and T.H.Gulliksen: Some reduction formulas for the Poincaré series of modules, Att. Acad. Naz. Lincei Rend. Cl. Sci. Fis. Natur. (8) 58 (1975), 82-91.

[30] E.Gover and M.Ramras: Increasing sequences of Betti numbers, Pacific J. Math. 87 (1980), 65-68.

[31] T.H.Gulliksen: A change of rings theorem, with applications to Poincaré series and intersection multiplicity, Math. Scand. 34 (1974), 167-183.

[32] T.H.Gulliksen: On the deviations of a local ring, Math. Scand. 47 (1980), 5-20.

[33] J.Herzog, B.Ulrich, and J.Backelin: Linear maximal Cohen-Macaulay modules over strict complete intersections, J. Pure Appl. Algebra 71 (1991), 187-201.

[34] A.Iarrobino: *Punctual Hilbert Schemes*, Memoirs Amer. Math. Soc. 188, AMS, Providence, RI, 1977 .

[35] C.Jacobsson: Finitely presented graded Lie algebras and homomorphisms of local rings, J. Pure Appl. Algebra 38 (1985), 243-253.

[36] C.Jacobsson, A.Kustin, and M.Miller: The Poincaré series of a codimension four Gorenstein ring is rational, J. Pure Appl. Algebra 38 (1985), 255-275.

[37] J. Lescot: Asymptotic properties of Betti numbers of modules over certain rings, J. Pure Appl. Algebra 38 (1985), 341-355.

[38] J.Lescot: Séries de Poincaré des modules multi-gradués sur les anneaux monomiaux, in: Lecture Notes in Math. 1318 (Springer-Verlag, Berlin, 1988), 155-161.

[39] J. Lescot: Séries de Poincaré et modules inertes, J. Algebra 132 (1990), 22-49.

[40] G.Levin: Poincaré series of modules over local rings, Proc. Amer. Math. Soc. 72 (1978), 6-10.

[41] C.Löfwall and J.-E.Roos: Cohomologie des algèbres de Lie graduées et séries de Poincaré-Betti non rationnelles, C. R. Acad. Sci. Paris 290 (1980), Ser. A, 733-736.

[42] S.Palmer: Algebra structures on resolutions of rings defined by grade four almost complete intersection ideals, Preprint, 1990.

[43] S.Priddy: Koszul resolutions, Trans. Amer. Math. Soc. 152 (1970), 39-60.

[44] M.Ramras: Betti numbers and reflexive modules, in: *Ring Theory* (Academic Press, New-York, 1972), 297-307.

[45] M.Ramras: Sequences of Betti numbers, J. Algebra 66 (1980), 193-204.

[46]. M.Ramras: Bounds on Betti numbers, Can. J. Math. 34 (1982), 589-592.

[47] J.-E.Roos: Homology of loop spaces and of local rings, in: *Proceedings 18th Scandinavian Congress of Mathematics*, 1980 (Birkhauser, Basel, 1981), 441-468.

[48] I.R.Shafarevich: Deformations of commutative algebras of class 2 , Algebra i Analiz 2 (1990), No. 6 , 178-194 . [in Russian]; English translation: Leningrad Math. J. 2 (1991), No. 6 (to appear).

[49] J.Tate: On the homology of Noetherian rings and of local rings, Illinois J. Math. 1 (1957), 14-27.

Note added in August, 1991. A series of recent preprints of Kustin and Kustin with Palmer has added to the list (*) the following condition:

(g) edim R − dept R = 4 , the ring R is Cohen–Macaulay and contains 1/2 .

Thus, the almost complete intersections of codimension 4 , whose residue field has characteristic different from 2 , provide a further class of local rings for which Problems 1, 2, 6, 7, 11, 15, and 20 have been solved.

PROBLEMS ON BETTI NUMBERS
OF
FINITE LENGTH MODULES

H. CHARALAMBOUS AND E. G. EVANS, JR.

Throughout this paper R will be a local ring with residue field k and dimension d.

In their article [B-E 1] Buchsbaum and Eisenbud showed that if the minimal resolution of an ideal I of finite colength over a regular local ring of dimension d has an associative multiplicative structure, then the rank of the ith syzygy of R/I is at least $\binom{d-1}{i-1}$. Similarly the ith betti number, β_i, is at least $\binom{d}{i}$. They conjectured that there always was a multiplication and that the ranks and betti numbers were always at least this big. Later Avramov in [Av] showed that one cannot have such multiplicative structures in general. In fact he showed that one cannot even have the minimal resolution of R/I be a module over the koszul complex on a system of parameters contained in I.

Our first two problems come directly from the conjectures of Buchsbaum and Eisenbud. The latter problems come from the literature or discussions with various mathematicians. Many of these discussions took place in the lively informal sessions at the Sundance conference. It has not always been possible to recall who made the original suggestions that led to the problems. The authors wish to thank the organizers of Sundance III for creating the possibility for these discussions.

Problem 1 If R is a regular local ring of dimension d and M is a module of finite length, is the rank of the ith syzygy at least $\binom{d-1}{i-1}$?

This is known (and fairly elementary) up through dimension 4. One can view this as a question about vector bundles on the punctured spectrum of a regular local ring. One can see Horrocks lovely article [H] or the book by Evans and Griffith [E-G 2] for details. If R is a regular local ring and M is a finitely generated module which is free on the punctured spectrum of a regular local ring and if $\mathcal{M}$ is its sheaf of sections, then $H^i(\mathcal{M})$ is isomorphic to $\text{Ext}^i_R(M^*, R)$ where M^* is $\text{Hom}(M, R)$. Using

Both authors were partially supported by the National Science Foundation .

Typeset by $\mathcal{A}\mathcal{M}\mathcal{S}$-TEX

25

the Ext criteria for a module to be a module of kth syzygies of a finite length module, the question asks if there is a module M such that $\mathcal{M}$ is a vector bundle on the punctured spectrum of R of rank less than $\binom{d-1}{k-1}$ with $H^i(\mathcal{M})$ equal to zero for all i except for $i = 0$ and $k-1$. Both H^0 and H^{k-1} are nonzero in any case.

Problem 2 If R is a regular local ring of dimension d and M is a module of finite length is the ith betti number at least $\binom{d}{i}$? This is also known and elementary up through dimension 4.

These two problems were submitted by G. Horrocks as part of R. Hartshorne's problem list [Ha]. This caused many authors to refer to them as Horrocks' questions or even Horrocks' conjectures. However the article by Buchsbaum and Eisenbud in which they make these conjectures occurs well prior to Hartshorne's list.

It is clear that an affirmative solution to problem 1 will give an affirmative solution to problem 2. For the special case of multigraded modules both problems are solved in the affirmative. This was done in a series of papers ([E-G 1], [San], and [C]). Huneke and Ulrich [H-U] showed that the answers to problem 1 is affirmative for R/I if I is in the linkage class of a complete intersection. The smallest possible counter example that might exist is in dimension 5 with betti numbers (1, 7, 9, 9, 7, 1). First one uses Kunz's theorem to show that the smallest number of generators for a Gorenstein that is not a complete intersection is 7. Then one uses that the rank of the third syzygy is at least three to show that the middle two betti numbers are at least 9. The smallest known betti sequences for height 5 Gorenstein ideals is (1, 7, 16, 16, 7, 1). Such an example arises by taking a 5 generated height three Gorenstein ideal and killing two new variables. For example the ideal $I = (a^2 + bc, b^2, c^2, ab, ac, d, e)$ will work. All known 7 generated height 5 Gorenstein ideals are in the linkage class of a complete intersection. It is conjectured that they all are. If they were, the result of Huneke and Ulrich mentioned above would show that their betti numbers would satisfy the conditions of problems one and two. Additional affirmative information for these problems can be found in the paper by Herzog and Kühl [H-K].

There is an interesting implication from problem 2 back to the stronger problem 1. Let $R = S[[x]]$ be a power series ring and M a finite length module over R. Then one can view M as a finite length module over S also. In [C] Charalambous showed that the rank of the ith syzygy of M over R is exactly the dimension over k of the cokernel of the map induced on $\text{Tor}_i^S(M, k)$ by the nilpotent S endomorphism coming from multiplication by x. Thus, if problem 2 were correct over S, then Tor would have dimension big enough and one could hope that the dimension of the cokernel is also appropriately big. This observation plays a crucial role in the case of multigraded modules.

There are a few fairly elementary observations that take care of the low dimensional cases. First, if M is a finite projective dimension finite length module over a ring R of dimension d, then $\text{Ext}^d(M, R)$ will be another such module whose resolution is exactly the dual of the original one. This means that one only needs to handle half the cases. Second if M is the cokernel of a map from R^n to R^m and the dimension of R is d, then $n - m + 1$ must be at least d in order that the height of the ideal generated by the maximal minors of the map be primary to the maximal ideal. Furthermore if one has equality in the above then the module is resolved by the generalized Buchsbaum-Rim complex, see [B-R] and [B-E 2] for details, and the rest of its betti numbers are known: $\beta_i = \binom{m+i-3}{i-2}\binom{n}{m+i-1}$. One can easily see that they satisfy the requirements of problems 1 and 2. Finally if R contains a field, then the syzygy theorem asserts that the rank of a nonfree kth syzygy be at least k. See [E-G 2] for details.

Problem 3 The obstruction in Avramov's paper [Av] to creating a module structure over a koszul complex on some system of parameters on the minimal resolution can be made zero by replacing the given system of parameters by sufficiently high powers. If one could find a sufficiently high power of the system of parameters to get such a module structure, then one would solve problems 1 and 2. This problem asks if given a finite length module, is there some system of parameters, $x_1, \ldots, x_d$, so that the minimal resolution can be made a module over the koszul complex on the x_i?. Even more explicitly can one do this in Avramov's example where both the ranks and betti numbers are big enough to allow it? Recently Heema Srinivasan has shown that there is no algebra structure on the resolution of $R[x_{ij}]$ modulo the ideal generated by the 4 by 4 Paffians of x_{ij} where x_{ij} is a 6 by 6 generic skew symmetric matrix even though Avramov's invariant vanishes in that case.

Problem 4 Although the original questions were over a regular local ring, no examples of a negative solution to problems 1 and 2 are known even if the module is not assumed to have finite projective dimension. Thus we can ask problem 1 (taking whatever definition of rank one wants in the non domain case) and problem 2 in the case that the local ring of dimension d is not assumed to be regular or the module to have finite projective dimension. Avramov and Buchweitz have generalized this still further in their unpublished talks from the 1987 Micro program on commutative algebra at MSRI. If one has a free complex over R, one can make a large matrix out of it by just taking the direct sum of the modules and considering the differential as an endomorphism of that module. This yields a lower triangular matrix of square zero with finite length homology. Thus given any local ring of dimension d and a lower triangular matrix of square zero and finite length homology, one can ask if the the size of the matrix is at least 2^d, if the matrix breaks into block lower triangular form and if the size of those blocks is at least the desired binomial coefficient.

Another generalization comes from Paul Roberts' Intersection Theorem [R]. If over any local ring one has a complex of finitely generated free modules with all homology of finite length, then his theorem shows that the length of the complex is at least the dimension of the ring. One can ask if in this setting the size of the free modules must be at least the desired binomial coefficients.

Carlsson has also formulated a version of these conjectures coming from the topological viewpoint. His primary interest is free actions of groups on products of spheres but he rephrases some of the material in purely algebraic terms. He has several formulations on differential graded modules similar to the situation above. He verifies these conjectures in low dimension. The interested reader should consult [Ca 1] and [Ca 2] for details.

Problem 5 If $m \in M$ is a minimal generator, then one can map M onto k carrying m to 1. m is killed by some maximal R sequence generating an ideal I. Thus one gets a map from R/I to k factoring through M. Lifting these maps to the resolutions one gets a map from the koszul complex on the generators of I to the koszul complex on the generators of the maximal ideal. This map factors through the resolution of M. If the ranks of the comparison maps between the koszul complex on the z_i and the koszul complexes resolving k had rank equal to the appropriate binomial coefficient, then the betti numbers of M would be equal to that binomial coefficient also. Unfortunately the ranks are not always that size. For example, in dimension 4 if $I = (a,\ b,\ c,\ d)$ and $J = (a^2,\ b^2,\ c^2,\ d^2)$, then

$$
\begin{array}{ccccccccccc}
0 \longrightarrow & R & \longrightarrow & R^4 & \longrightarrow & R^6 & \longrightarrow & R^4 & \longrightarrow & R & \longrightarrow & R/J \longrightarrow 0 \\
& \downarrow{\scriptstyle m.4} & & \downarrow{\scriptstyle m.3} & & \downarrow{\scriptstyle m.2} & & \downarrow{\scriptstyle m.1} & & \downarrow{\scriptstyle m.0} & & \downarrow \\
0 \longrightarrow & R & \longrightarrow & R^4 & \longrightarrow & R^6 & \longrightarrow & R^4 & \longrightarrow & R & \longrightarrow & R/I \longrightarrow 0
\end{array}
$$

Note that $m.0 = 1$ the identity map on R. An obvious choice for $m.1$ is the map represented by the matrix

$$
\begin{pmatrix}
a & 0 & 0 & 0 \\
0 & b & 0 & 0 \\
0 & 0 & c & 0 \\
0 & 0 & 0 & d
\end{pmatrix}.
$$

The following two matrices are posssible choices for $m.2$:

$$\begin{pmatrix} ad & 0 & 0 & 0 & 0 & 0 \\ 0 & bd & 0 & 0 & 0 & 0 \\ 0 & 0 & cd & 0 & 0 & 0 \\ 0 & 0 & 0 & ac & 0 & 0 \\ 0 & 0 & 0 & 0 & bc & 0 \\ 0 & 0 & 0 & 0 & 0 & ab \end{pmatrix} \quad \begin{pmatrix} ad & -c^2 & 0 & -b^2 & 0 & 0 \\ 0 & bd & 0 & ab & 0 & 0 \\ 0 & ac & cd & 0 & 0 & 0 \\ 0 & cd & 0 & ac & 0 & 0 \\ 0 & 0 & 0 & 0 & bc & 0 \\ 0 & 0 & 0 & bd & 0 & ab \end{pmatrix}$$

It is easy to verify that the first matrix (which is the second exterior power of $m.1$) has rank 6 while the second one (which adds appropriate multiples of the map from R^4 to R^6 in the resolution of R/I)) has rank only 5. The question is what is the smallest possible rank of such comparison maps between koszul complexes in dimension d.

Problem 6 A similar approach for the cyclic case would be to take a system of parameters inside the ideal and look at the resulting comparison maps on the resolutions. To make the problem more precise, suppose that $I = (f_1, \ldots, f_n)$ and (a_{ij}) is a d by n matrix of indeterminates over R and that $J = (\Sigma_{j=1}^n a_{1j} f_j, \ldots, \Sigma_{j=1}^n a_{dj} f_j)$, then is some (every) comparison map from the resolution of $R[a_{ij}]/J$ to the resolution of $R[a_{ij}]/I$ a monomorphism? We remark that this situation arises in the generic linkage discussed in [H-U].

Problem 7 If problem 2 was solved in the affirmative then it would follow that $\Sigma_{i=0}^d \beta_i \geq 2^d$. Thus a weaker question asks if this is true. We remark that the possible counter example to problems 1 and 2 in dimension 5 still has $\Sigma_{i=0}^d \beta_i = 32 = 2^5$. Indeed in unpublished conversations Avramov and Buchweitz have shown that for all modules of finite length in dimension 5 over an equicharacteristic regular local ring, $\Sigma_{i=0}^5 \beta_i \geq 32$. Also in unpublished conversations Stanley, Avramov, and Buchweitz have shown that if R is a graded polynomial ring over a field and M is a graded module of odd length or more generally, if the alternaring sum of the graded pieces of M is nonzero then $\Sigma_{i=0}^d \beta_i \geq 2^d$. The first unknown case is in dimension 6. Using only duality, the bound on the heights of determinental ideals, the syzygy theorem, and Kunz's [K] theorem that a Gorenstein ideal cannot be an almost complete intersection, one cannot rule out the existence of a finite length module in dimension six with betti sequence (1, 8, 11, 8, 11, 8, 1) whose sum is just 48 rather than the desired 64. The existence of such a sequence seems rather unlikely.

Problem 8 The central idea in the result of Stanley, Avramov, and Buchweitz mentioned above is a rather simple method for estimating the sum of the betti numbers from the Hilbert series of a module. In particular if M is a finite length module, then the Hilbert series of M is just a polynomial with non negative integer coefficients. We reproduce their idea

here. One can (as Hilbert did for the Hilbert function in his 1890 paper)
read the Hilbert series from the graded resolution as the alternating sum
of the Hilbert series of the free modules. Note that the Hilbert series of
$R[-n]$ where all the variables are given the degree 1 is $t^n/(1-t)^d$. Let M
be the module in question and

$$0 \longrightarrow R[-n_{d1}] \oplus \cdots \oplus R[-n_{dr}] \longrightarrow \cdots \longrightarrow R[-n_{11}] \oplus \cdots \oplus R[-n_{1s}]$$
$$\longrightarrow R[-n_{01}] \oplus \cdots \oplus R[-n_{0l}] \longrightarrow M \longrightarrow 0$$

be a graded free resolution of M. We remark that with the above notation
$l = \beta_0(M)$, $s = \beta_1(M)$ and so on. The Hilbert series of M,

$$H_M(t) = \sum \text{length}(M_i)t^i,$$

is equal to

$$H_M(t) = \frac{(t^{n_{01}} + \cdots + t^{n_{0l}}) - \cdots + (-1)^d(t^{n_{d1}} + \cdots + t^{n_{dr}})}{(1-t)^d}.$$

Note that since $H_M(t)$ is a polynomial the equality $H_M(t)(1-t)^d = (t^{n_{01}} + \cdots + t^{n_{0l}}) - (t^{n_{11}} + \cdots + t^{n_{1s}}) + \cdots + (-1)^d(t^{n_{d1}} + \cdots + t^{n_{dr}})$ makes sense
for all numbers t. Now to get the desired bound on the sum of the betti
numbers of M we only have to evaluate this expression at $t = -1$ and take
absolute values. On the left side we get the alternating sum of the lengths
of the M_i times 2^d and on the right side we get a sum less than or equal
to the sum of the betti numbers. If the length of M is odd, the alternating
sum of the lengths of the M_i cannot be zero and thus the sum of the betti
numbers must be at least 2^d.

More generally given a Hilbert series one can always get a module with
that Hilbert series by taking a direct sum of the appropriate number of
copies of the residue field k in each degree. The resolution of such a module would look like a direct sum of copies of the koszul complex on the
generators of the maximal ideal suitably regraded. Any other resolution
with the same Hilbert series must have the same graded terms except that
if $R[-n]$ occurs in both an odd and an even part of the resolution one
can remove both of these terms. After having done this as many times as
possible, one gets certain number of copies of $R[-n]$ that must occur in the
odd positions for some n and certain $R[-m]$ that must occur in the even
positions. Often this information is enough to settle that $\Sigma_{i=0}^d \beta_i \geq 2^d$.
Alas not always. For example, if $I = (a,\ b,\ c^2)$, the resolution of R/I is

$$0 \longrightarrow R[-4] \longrightarrow R[-2] \oplus R[-3] \oplus R[-3] \longrightarrow R[-1] \oplus R[-1] \oplus R[-2]$$
$$\longrightarrow R \longrightarrow R/(a, b, c^2) \longrightarrow 0.$$

The generator c^2 gives a copy of $R[-2]$ in the first syzygies while the koszul relation on a, b gives a copy of $R[-2]$ in the second syzygies and these two copies cancel each other in the alternating sum. More explicitly this resolution shows that the Hilbert series for R/I, $1+t$, can be written as

$$\frac{1 - 2t - t^2 + t^2 + 2t^3 - t^4}{(1-t)^3} = \frac{1 - 2t + 2t^3 - t^4}{(1-t)^3}$$

which needs only six terms. Thus this argument only shows that $\Sigma_{i=0}^{d} \beta_i \geq$ 6. The problem is for a given Hilbert series, what are the minimal possible betti numbers for modules that achieve that series? The direct sum of copies of k gives the largest betti numbers. Somewhat more specifically, if the Hilbert series is the Hilbert series of a cyclic module (criteria for this have been known since the time of Macaulay, see Stanley's book [St] for a modern treatment), what are the minimal possible betti numbers for a cyclic module that achieve that Hilbert series? The maximal possible ones are discussed in unpublished work of Stillman et al.

Problem 9 Lots of examples of the resolution of finite length modules have been obtained by the authors (and many others) using the computer program Macaulay. In all of these cases except for R/I where I is generated by an R sequence the betti numbers are much bigger than required by problem 2. Of course if the module is R modulo a maximal R sequence then the betti numbers are exactly what is asked for. We partially order the betti sequences by $(\beta_0, \ldots, \beta_d) \geq (\alpha_0, \ldots, \alpha_d)$ if $\beta_i \geq \alpha_i$ for all i. This problem asks what are the minimal possible betti sequences for the class of modules of finite length which are not isomorphic to R modulo a maximal sequence? It follows from the fact that polynomial rings are noetherian that there are only finitely many minimal sequences. They are explicitly known for dimension up to three. For dimension 4 one knows six sequences such that every betti sequence is bigger than at least one of that six but two of the sequences are not known whether they correspond to modules or not. The unknown sequences are (1, 6, 9, 6, 2) and its dual (2, 6, 9, 6, 1). The other minimal ones are (2, 6, 8, 6, 2), (1, 6, 10, 6, 1), (1, 5, 9, 7, 2), and its dual (2, 7, 9, 5, 1). In dimension 5 there is no substantial guess as to what the minmal sequences are. If the unknown ones in dimension 4 do not in fact exist, one could reasonably guess that the ones that occur in dimension 5 come from taking modules in dimension 4 with minimal sequences and letting a new variable act on them by zero. One notes that the minimal sequences that are known to exist in dimension 4 come from ones in dimension three in this fashion. See [C-E-M] for details. It is interesting to note that the crucial information in the arguments in this paper come from the knowledge of the algebra structures on Gorenstein ideals in 4 variables provided by [K-M].

Problem 10 By looking at the sequences in [C-E-M] one can see that up through dimension 4 if M is not R modulo a maximal sequence then $\Sigma_{i=0}^{d}\beta_i \geq 2^d + 2^{d-1}$ and that $\beta_i > \binom{d}{i}$ for all i except 0 and d. One can ask if this always happens. In addition to the results for dimension up through 4, this is also known for all finite length multigraded modules [C].

Problem 11 There are several questions that need addressing that cannot be made into such explicit problems but may be worth thinking about anyway. The Hilbert polynomial or series gives one useful way of condensing the information in the resolution. Chern polynomials give yet another. Horrocks [H] uses them in this area as do many others. Can further use be made of them to get estimates on the betti numbers? Gorenstein ideals are reasonably well understood thanks to the work of Buchsbaum, Eisenbud, Kustin, Miller, Huneke, Ulrich, and many others. Can one say anything more about Cohen-Macaulay ideals of type 2 (i.e. the highest betti number is 2)? Thanks to the understanding of Gorenstein ideals and linkage, we know a lot about almost complete intersections. Can we say anything about Cohen-Macaulay ideals of deviation (the number of generators minus the height) 2? Of course a particular question for both of these problems is the existence of an ideal I in a 4 dimensional ring R so that R/I has betti sequence (1, 6, 9, 6, 2). We remark that if I is an ideal of deviation 2 and we form its link, J, using as the linking R sequence a subset of minimal generators of I, then J will be an ideal of type 2. Thus these last two questions are "linked".

REFERENCES

[Av]   L. Avramov, *Obstructions to the existence of multiplicative structures on minimal free resolutions*, American J. Math. **103** (1981), 1–31.

[B-E 1]   D. Buchsbaum and D. Eisenbud, *Algebra structures for finite free resolutions and some structure theorems for ideals of codimension 3*, American J. of Mathematics **99**, No 3 (1977), 447–485.

[B-E 2]   D. Buchsbaum and D. Eisenbud, *Generic free resolutions and a family of generically perfect ideals*, Adv. in Math. **18** (1975), 245–301.

[B-R]   D. Buchsbaum and D. S. Rim, *A generalized Koszul complex, II: depth and multiplicity*, Trans. Amer. Math. Soc. **111** (1964), 197–224.

[Ca 1]   G. Carlsson, *Free $(Z/2)^k$-actions and a problem in commutative algebra*, Lecture Notes in Mathematics, Springer-Verlag **1217**, 79–83.

[Ca 2]   G. Carlsson, *Free $(Z/2)^3$-actions on finite complexes*, Annals of Mathematical Studies **113** (1983), 332–344.

[C]   H. Charalambous, *Lower bounds for betti numbers of multigraded modules*, J. Algebra **137** (1991), 491–500.

[C-E-M]   H. Charalambous, E. G. Evans, and M. Miller, *Betti numbers for modules of finite length*, Proc. of the A. M. S. **109** (1990), 63–70.

[E-G 1]   E.G. Evans and P. Griffith, *Binomial behavior of betti numbers for modules of finite length*, Pacific J. of Mathematics **133** (1988), 267–276.

[E-G 2]   E.G. Evans and P. Griffith, *Syzygies*, Cambridge University Press London Society Lecture Notes 106, Cambridge, 1985.

[Ha]   R. Hartshorne, *Algebraic vector bundles on projective spaces: a problem list*, Topology 18 (1979), 117-128.

[H-K]   J. Herzog and M. Kühl, *On the betti numbers of finite pure and linear resolutions*, Comm. Alg. 12 13-14 (1984), 1627-1646.

[H]   G. Horrocks, *Vector bundles on the punctured spectrum of a regular local ring*, Proc. London Math. Soc. (3) 14 (1964), 689-713.

[H-U]   C. Huneke and B. Ulrich, *The structure of linkage*, Annals of Mathematics 126 (1987), 277-334.

[K]   E. Kunz, *Almost complete intersections are not Gorenstein rings*, J. Algebra 28 (1974), 111-115.

[K-M]   A. Kustin and M. Miller, *Classification of the Tor-algebras of codimension four Gorenstein local rings*, Math. Z. 190 (1985), 341-355.

[R]   P. Roberts, *Le Théorème d'intersection*, C. R. Acad. Sci. Paris Sér 304 No. 7 (1987), 177-180.

[San]   L. Santoni, *Horrocks' Question for monomially graded modules*, Pacific J. of Mathematics 141, No 1 (1990), 105-124.

[St]   R. Stanley, *Combinatorics and Commutative Algebra*, Birkhaüser, Boston, 1983.

DEPARTMENT OF MATHEMATICS, UNIVERSITY OF ILLINOIS, URBANA, IL 61801

MULTIPLICATIVE STRUCTURES
ON FINITE FREE RESOLUTIONS

MATTHEW MILLER

We begin with a brief recapitulation of the definitions. Following this we intersperse open problems, questions, and conjectures with summaries of the known results in three categories: existence results, non-existence results (obstructions), and classification results. Throughout $R, \mathfrak{m}, k$ denotes a local ring (or, if the obvious modifications are made, a graded ring of form $\oplus R_n$ with $R_0 = k$, $\mathfrak{m} = R_+$, and $R = R_0[R_1]$), all modules are finitely generated, and all resolutions, unless otherwise indicated, are minimal finite free resolutions of cyclic modules. By a Gorenstein ideal we mean a perfect ideal I of grade g with $\operatorname{Ext}_R^g(R/I, R) \cong R/I$. The ideal generated by the $m \times m$ minors of a matrix X is written $I_m(X)$. Standard references for linkage theory are [6] and [17]. An ideal is *licci* if it is in the linkage class of a complete intersection.

1. EXISTENCE RESULTS

A resolution $\mathbb{F}, d$

$$0 \to F_g \xrightarrow{d_g} F_{g-1} \xrightarrow{d_{g-1}} \ldots \xrightarrow{d_2} F_1 \xrightarrow{d_1} F_0 = R$$

of a cyclic module R/I is called an *algebra resolution* if there is an associative multiplication (called a DG-algebra structure) $\mathbb{F} \otimes \mathbb{F} \to \mathbb{F}$ extending the R-module structure of F_i and satisfying

- $F_i F_j \subset F_{i+j}$

- $x_i x_j = (-1)^{i+j} x_j x_i$ for $x_i \in F_i$ and $x_j \in F_j$ (and $x_i^2 = 0$ if i is odd)
- $d(x_i x_j) = (dx_i)x_j + (-1)^i x_i dx_j$ for $x_i \in F_i$ and $x_j \in F_j$

If also $d\mathbb{F} \subset \mathfrak{m}\mathbb{F}$ then we call $\mathbb{F}$ a *minimal* algebra resolution. The best known example, of course, is the Koszul resolution $\mathbb{K}$ of $R/(\mathbf{a})$ for a regular sequence $\mathbf{a} = a_1, \ldots, a_g$; this has the structure of an exterior algebra on R^g. The other known existence results are the following:

- $\mathbb{F}$ has length $g \leq 3$ (see [6])
- $\mathbb{F}$ has length $g = 4$, I is a Gorenstein ideal (see [12] and [9])
- $\mathbb{F}$ has length $g \geq 3$, I is a Herzog ideal of grade g, i.e., a Gorenstein ideal of grade g linked in two steps to a complete intersection (see [14])
- $\mathbb{F}$ has length $g \geq 3$, I is a Northcott ideal, i.e., an almost complete intersection of grade g directly linked to a complete intersection (see [5])
- $\mathbb{F}$ is the Eagon-Northcott resolution of R/I for $I = I_m(X_{n \times m})$, the ideal of maximal minors of a generic matrix, assuming that R contains the rationals; also the resolution of $R/(\mathbf{a})^n$ for $n > 1$ and $\mathbf{a}$ a regular sequence in R (see [18])
- $\mathbb{F}$ has length $g \geq 5$ (odd) and I is the Gorenstein ideal of the grade g, deviation 2 (i.e., $\mu(I) = g + 2$), defined by Huneke and Ulrich (see [19])
- $\mathbb{F}$ has length $g = 4$, I is an almost complete intersection (see [16] and [11]).

While all the proofs of these results are computational, they have in fact two significantly different flavors: an associative product is constructed by modification of a non-explicit first approximation (product structures that are not necessarily associative are easily shown to exist) in the first, second, and last of the results listed above; explicit determination of the multiplication tables is accomplished for the others. In light of many of the above results (the Huneke-Ulrich ideals are also linked to a complete intersection; in grade g this takes $g - 1$ steps), a positive solution to the following problem seems likely.

Problem 1. If I is in the linkage class of a complete intersection, determine whether the minimal free resolution of R/I is a DG-algebra.

From a computational point of view it would probably be best to begin with the almost complete intersections that form the intermediate steps in the standard chain of links from a Huneke-Ulrich ideal to a complete intersection. On the other hand one might hope for an enormous generalization of the existence of minimal algebra resolutions for Herzog ideals. Ulrich [21] has shown that if I and J are geometrically linked licci ideals of grade $g > 0$, then $K = I + J$ is also licci (and Gorenstein of grade $g + 1$ by [17]); and conversely, if K is a licci Gorenstein ideal of grade at least two, then K is a "generalized localization" of such a sum. Herzog ideals are constructed by taking I to be a complete intersection and K to be a deformation of $I + J$ (this is included in "generalized localization").

Problem 2. Suppose I and J are geometrically linked licci ideals and R/I has a minimal algebra resolution. Let K be a "generalized localization" of $I + J$. Determine whether R/K has a minimal algebra resolution.

Since DG-algebra structures fail to exist on minimal resolutions in general, even for Gorenstein ideals, but are easily seen to exist on sufficiently non-minimal resolutions (using Tate's construction), it is reasonable to ask whether one can pare down a non-minimal resolution and still retain the algebra structure. (Indeed this part of the argument in [14], [11], and [19].) For some purposes non-minimal algebra resolutions may actually be "good enough." In the Gorenstein case, one would like to obtain a resolution that at least exhibits the usual symmetry. Avramov has posed the following problem.

Problem 3. If I is a Gorenstein ideal of grade g in R, does there exist an R-free algebra resolution $\mathbb{F}$ of length g such that

$F_g \cong R$ and $\mathbb{F}$ has Poincaré duality? The first open case is in grade 5.

The results of Palmer and Kustin point in a similar direction. They produce algebra resolutions by a mapping cone construction, which, while not necessarily minimal are not wildly non-minimal. Further, Kustin [10] has classified the possible structures of the k-algebra $\mathrm{Tor}_\bullet^R(R/I, k)$ for grade four almost complete intersections I by using these non-minimal algebra resolutions.

2. OBSTRUCTIONS (NON-EXISTENCE RESULTS)

On the other hand, definitive non-existence results have been found. There are examples of Hinič [1], reinterpreted by Avramov, of modules with no minimal algebra resolution. Avramov [2] has defined an obstruction $o^f(M)$, for M an S-module and $f \colon R \to S$ a small local homomorphism, to be the homology:

$$H(\mathrm{Tor}_+^R(S, k) \otimes \mathrm{Tor}^R(M, k) \xrightarrow{\ \mathrm{mult}\ } \mathrm{Tor}^R(M, k)$$

$$\xrightarrow{\ \mathrm{Tor}(f)\ } \mathrm{Tor}^S(M, k)).$$

If $o^f(M) \neq 0$, then the minimal R-free resolution $\mathbb{F}$ of M does not even have a module structure over any minimal R-free resolution $\mathbb{P}$ of S that has a DG-algebra structure. This is usually exhibited by taking $M = R/I$, **a** a regular sequence in I, $S = R/(\mathbf{a})$, and $\mathbb{P}$ the Koszul resolution of S; we shall denote the obstruction $o^{\mathbf{a}}(M)$ in this case. Avramov showed that if depth $R \geq n \geq m \geq 2$ and $n \geq 4$, then there exists a perfect ideal I of grade n and a regular sequence **a** in I of length m so that $o^{\mathbf{a}}(R/I) \neq 0$; if $n \geq 6$, then I can even be taken to be a Gorenstein ideal.

As a preliminary step on the way one would like to have a better idea how persistent the obstructions defined by Avramov might be.

Problem 4. If $o^f(R/I) = 0$ for all f that factor $R \to R/I$ through a complete intersection, is this still true for R/J in the linkage class of R/I?

The vanishing of these obstruction modules for all complete intersection maps f does not, however, guarantee existence of a minimal algebra resolution. In a recent preprint, Srinivasan [20] has exhibited a Gorenstein ideal I with $o^{\mathbf{a}}(R/I) = 0$ for every regular sequence $\mathbf{a}$ in I, but R/I nevertheless does not admit a minimal algebra resolution. In her example I is generated by the 4×4 pfaffians of a generic 6×6 alternating matrix.

Problem 5. Determine whether the minimal resolutions of all lower order pfaffian ideals lack algebra structure.

Problem 6. Develop a general theory of obstruction that would extend and account for Srinivasan's example.

It can happen that $o^{\mathbf{a}}(M) \neq 0$ for some regular sequence $\mathbf{a}$, so M has no minimal algebra resolution; but if $\mathbf{b}$ is a regular sequence contained in $\mathfrak{m} \cdot \mathrm{ann}\, M$, then $o^{\mathbf{b}}(M) = 0$ and the possibility of a DG-module structure remains. In particular if each element a_i of the regular sequence $\mathbf{a}$ is replaced by a_i^2, then the obstruction vanishes.

Problem 7. (See [2, 6.5.1]) What are the obstructions to the existence of $\mathbb{K}(\mathbf{a})$-module structure on a minimal resolution of M if $\mathbf{a} \subset \mathfrak{m} \cdot \mathrm{ann}\, M$?

One of the original motivations for producing multiplicative structures was the observation by Buchsbaum and Eisenbud [6] that the lower bound estimate for betti numbers of a perfect cyclic module M of codimension c, namely $b_i(M) \geq \binom{c}{i}$, follows from existence of a minimal algebra resolution $\mathbb{F}$ of M. Essentially a comparison map from a Koszul resolution to $\mathbb{F}$ is forced to be an injection (see the article by Charalambous and Evans [7]). Avramov [2] showed how to modify the argument if only a $\mathbb{K}(\mathbf{a})$-module structure exists; M need not even be cyclic.

Problem 8. If $M = R/I$ is a perfect module of codimension c does there exist a regular sequence $\mathbf{a} \subset I$ and an injection

$\mathbb{K}(\mathbf{a}) \rightarrow \mathbb{F}$?

3. CLASSIFICATION RESULTS

Throughout this section $S = R/I$ is a cyclic module with a finite free resolution $\mathbb{F}$ over a local ring $R, \mathfrak{m}, k$. All the known algebra structures on such resolutions look very complicated, but the induced k-algebra structure on homology

$$T = \operatorname{Tor}_{\bullet}^R(S, k) = H_{\bullet}(\mathbb{F} \otimes_R k) = \mathbb{F}_{\bullet} \otimes_R k,$$

which is unique (and exists even if $\mathbb{F}$ itself has no DG-algebra structure), is always very simple. For perfect ideals of grade three (also for imperfect ideals with $\operatorname{pd} S = 3$ if R is regular) there are five distinct types [22], [5]; for Gorenstein ideals of grade four there are four distinct types [15]; for perfect 5-generated ideals of grade four there are twelve types [10]. Some of these types depend on one or two integer parameters, typically $p = \dim T_1^2$ and $q = \dim T_1 T_2$. But, for instance, in the case of a length three resolution with T of form $\mathbf{G}(r)$ in the terminology of [5] (or $A_{0,r}^*$ in the terminology of [22]), r is given by the rank of the induced map $T_2 \rightarrow T_1^* \otimes T_3$.

Problem 9. Classify $T = \operatorname{Tor}_{\bullet}^R(S, k)$ for (perfect) ideals of grade $g \geq 4$, if possible without invoking the existence of minimal algebra resolutions. For $g = 4$ determine if (as in the Gorenstein and almost complete intersection cases) there are only finitely many types and if so, what are the critical numerical parameters. If (as seems likely) for $g \geq 4$ no such classification is possible, find appropriate (possibly "continuous") parameters, or indeed even moduli spaces, for the possible structures. Interpret possible degenerations of structures.

Problem 10. In the situations that a classification type depends on a numerical parameter, determine if ideals exist that realize this type and all possible parameter values; in particular determine if the number of generators of an ideal is incompatible with certain structures on T.

One approach to these problems would be to extend Weyman's [22, Section 4] use of invariant theory from codimension three to higher codimensions. Existence of certain structure maps enables one to deduce that there are relations satisfied by any possible product structures on T. For example, in Weyman's case there is a commutative diagram

$$
\begin{array}{ccccc}
\bigwedge^2 F_3 & \longrightarrow & F_3 \otimes F_2 & \longrightarrow & S_2 F_2 \\
\bigwedge^2 c \uparrow & \overset{q}{\nwarrow} & bc \uparrow & \overset{p}{\nwarrow} & \uparrow S_2 b \\
D_2 F_2 \otimes \bigwedge^2 F_1 & \longrightarrow & F_2 \otimes \bigwedge^3 F_1 & \longrightarrow & \bigwedge^4 F_1
\end{array}
$$

in which the top maps are induced by d_3, the bottom maps are induced by d_2 (one needs D_2 rather than S_2 for the argument to work in characteristic two), and $b \colon \bigwedge^2 F_1 \to F_2$ and $c \colon F_2 \otimes F_1 \to F_3$ are product maps on $\mathbb{F}$. The vertical maps are induced by induced by b and c after applying appropriate symmetrization (the description of bc in [22] is lacking a component $F_2 \otimes \bigwedge^3 F_1 \overset{\tau}{\to} \bigwedge^3 F_1 \otimes F_2 \overset{\mu \otimes \mathrm{id}}{\longrightarrow} F_3 \otimes F_2$, where $\mu(x \wedge y \wedge z) = c(x \wedge b(y \wedge z))$; but this vanishes modulo $\mathfrak{m}$ if S is a not a complete intersection, so the rest of Weyman's argument is not affected); the homotopy maps p and q satisfy the obvious conditions. Existence of the map p and minimality of $\mathbb{F}$ imply that $S_2 \bar{b} = 0$ (bar denotes reduction modulo $\mathfrak{m}$); similarly existence of q and minimality of $\mathbb{F}$ imply $\bigwedge^2 \bar{c}$ and $\bar{b}\bar{c}$ are also zero modulo $\mathfrak{m}$. Hence there is a coordinate ring for the space of pairs $(\bar{b}, \bar{c})$ with an action of the group $G = GL(T_3) \times GL(T_2) \times GL(T_1)$. The orbits of G can then be described and classified. The "square zero" relation $S_2 \bar{b} = 0$ leads to a strong decomposability condition on $\bar{b}^*$, namely that each element of im $\bar{b}^*$ is decomposable in $\bigwedge^2 T_1^*$. (Another way to say this is that $\bar{b}^*(\phi)$ is an alternating matrix of rank at most two for every ϕ in T_2^*; see problem 14 for a possible generalization.) This severely constrains the possibilities for $\bar{b}$, and the vanishing of $\bar{b}\bar{c}$ and $\bigwedge^2 \bar{c}$ limits the possibilities for $\bar{c}$, given the form of $\bar{b}$. Notice that this approach does not involve associativity of the multiplication on $\mathbb{F}$ itself, though of course associativity of T plays a role.

One would like to deduce consequences for the structure of I from the structure of T (certainly this is one aspect of Problem 10). Very few such results are known, and these are mostly in small codimension. Avramov [3, Proposition 3.4] has shown that if either pd $S \leq 3$ or pd $S = 4$ and I is Gorenstein, then $I = (J, x)$ for some x regular on R/J if and only if T is free as a module over a subalgebra $k\langle e\rangle$, with deg $e = 1$. For arbitrary codimension Avramov and Foxby [4] have shown that if T is a Poincaré duality algebra, then I is a Gorenstein ideal (one need not even assume perfection of S).

Problem 11. If T is free as a module over a subalgebra $k\langle e\rangle$, with deg $e = 1$, then does I decompose as (J, x) with x regular on R/J?

There is some evidence that if I is a Gorenstein ideal, then the subspace spanned by the multiplication on T can not be too large, unless I is a complete intersection (this is purposely vague!). It is not even clear if the appropriate measure is absolute size (given the grade g or the number of generators of I) or perhaps relative size (how much of T_2, say, is spanned by T_1^2).

Problem 12. Find upper bounds for $\dim T_i T_j$ and for $\dim T_1^n$ if I is a Gorenstein ideal of grade $g \geq 5$, but not a complete intersection.

There are very few results of this sort in the literature. A trivial argument shows that I is a complete intersection if $T_1^g \neq 0$. In grade four there is the simple result [13] that $\dim T_1^2 > (1/2) \dim T_2$ implies I is a complete intersection; a quick glance at the classification theorem in [15] shows that $\dim T_1^2 = \dim T_1 T_2 \leq 3$ unless there is a distinguished element e in T_1 so that $T_1^2 = eT_1$ (and in any case $\dim T_1^3 = 0$). In grade $g \geq 5$ the Tor-algebra of the Herzog ideal has $\dim T_1^n = \binom{g-1}{n}$ for $n < g - 1$ and $T_1^{g-1} = 0$ (see [5, Proposition 6.3 and

Example 5.11]). It is possible that the Huneke-Ulrich ideals exhibit the maximum possible non-trivial multiplication in T for ideals that are not complete intersections: if the grade is five, then $\dim T_1 = \dim T_4 = 7$, $\dim T_2 = \dim T_3 = 22$, $\dim T_1^2 = \binom{6}{2} = 15$, $\dim T_1^3 = 0$, $\dim T_1 T_2 = 7$, $\dim T_1 T_3 = 6$, and $\dim T_2^2$ is either 6 or 7 according as 2 is a unit in R or not. If $g = 2n - 1$ (and I is therefore generated by $2n + 1$ elements), then the multiplication on T (see [19] or [8, Theorem 6.2]) satisfies $\dim T_1^i = \binom{2n}{i} = \binom{g+1}{i}$ for $i < n$ and $T_1^n = 0$.

After the Gorenstein ideals the next best class (certainly from a linkage-theoretic point of view) are the perfect almost complete intersections. In grade four the ideals of Palmer's thesis have $T = E' \ltimes W$, where $E' = E/E_4$ is a truncation of the exterior algebra $E = \bigwedge k^4$ and W is a trivial E'-module. In particular, $\dim T_1^i = \binom{4}{i}$ for $i \leq 3$. The Tor-algebras of the Northcott ideals given in [5] reveal the same structure, with $\dim T_1^i = \binom{g}{i}$ for $i \leq g - 1$ and $T_1^g = 0$. The Huneke-Ulrich almost complete intersections exhibit the phenomenon of having a large number of higher order Koszul relations in their minimal resolutions in an exceptionally strong way: in grade $g = 2n$ the Tor-algebra satisfies $\dim T_1^i = \binom{2n+1}{i} = \binom{g+1}{i}$ for $i \leq n$ and $\dim T_1^{n+1} = 0$ (see [8]), that is, the Koszul relations on *all* the minimal generators appear in the resolution as far as half-way back! In this case, however, we don't know the full structure of $\mathrm{Tor}_\bullet^R(S, k)$. Huneke has pointed out that by taking the product $S_1 \otimes_k S_2$ of two such (as a quotient of the product of disjoint polynomial rings), one obtains a perfect ideal of deviation two and grade $4n$ with $\dim T_1^2 = \binom{4n+2}{2}$. This process can be iterated so that one can produce perfect ideals of arbitrarily high deviation (but also with correspondingly high grade) such that $\dim T_1^2 = \binom{\mu(I)}{2}$. In Kustin's recent preprint [10] one finds grade four almost complete intersections with $6 \leq \dim T_1^2 \leq 10 = \binom{5}{2}$, $\dim T_1^3 = 10 - \dim T_1^2$, and various possibilities for $\dim T_1 T_j$.

Problem 13. For which ideals is $\dim T_1^2 = \binom{\mu(I)}{2}$?

If we consider the classification problem from the perspective of Weyman's analysis, then the same computational evidence points to the following problem. Here the emphasis is less on the size of T_1^2, but rather on how "decomposable" the product structure must be.

Problem 14. The image of the map $\bar{b}^* \colon T_2^* \to \bigwedge^2 T_1^*$ lies in the space of $n \times n$ alternating matrices, where $n = \dim T_1$. Excluding the examples $S_1 \otimes_k S_2$ mentioned above, determine whether each image $\bar{b}^*(\phi)$ is a matrix of rank at most $g + 1$. (In fact, in all known cases, except for the Huneke-Ulrich Gorenstein ideals of deviation two, one can replace $g + 1$ by g; and since the rank of an alternating matrix is even, one can lower an odd upper bound by one automatically.)

Finally, Weyman has suggested that one consider how T is generated as a k-algebra (intending application to resolutions of coordinate rings of Grassmannians.) At one extreme we have the exterior algebra structure $T = \bigwedge T_1$ for a complete intersection. On the other hand there are resolutions with long linear strands, so that generators of T must appear in high homological degrees. Suppose that R/I is a quotient of a polynomial ring $R = k[X_1, \ldots, X_n]$ by a homogeneous ideal. Then there is a graded minimal resolution and this induces a bigraded structure $T_{i,j} = \mathrm{Tor}_i^R(R/I, k)_j$. If I is generated by quadrics and the resolution of R/I has a linear strand, say $T_{i,i+1} \neq 0$ for $1 \leq i \leq p$, then T is generated over T_0 by this linear strand in many, but not all, examples (and, indeed, in all cases for which $p \geq 2$).

Problem 15. Find necessary and sufficient conditions for the linear strand in a resolution of an ideal generated by quadrics to generate T.

ACKNOWLEDGMENTS. Thanks to the organizers and participants of the conference at Sundance for stimulating a fresh look at this part of the theory of resolutions, and especially for posing new problems! Thanks also to Andy Kustin for very helpful discussions during the preparation of this article.

REFERENCES

1. L. Avramov, *On the Hopf algebra of a local ring*, Appendix by V. Hiníč, Izv. Akad. Nauk. SSSR Ser. Mat. **38** (1974), 253–277, English translation, Math. USSR-Izv. **8** (1974), 259–284.

2. L. Avramov, *Obstructions to the existence of multiplicative structures on minimal free resolutions*, Amer. J. Math. **103** (1981), 1–31.

3. L. Avramov, *Homological asymptotics of modules over local rings*, Commutative Algebra, Math. Sci. Research Institute Publications **15**, Springer Verlag, 1989, pp. 33–62.

4. L. Avramov and H.-B. Foxby, *Gorenstein local homomorphisms*, Bulletin Amer. Math. Soc. (1990), 145–150.

5. L. Avramov, A. Kustin, and M. Miller, *Poincaré series of modules over local rings of small embedding codepth or small linking number*, J. Algebra **118** (1988), 162–204.

6. D. Buchsbaum and D. Eisenbud, *Algebra structures for finite free resolutions, and some structure theorems for ideals of codimension 3*, Amer. J. Math. **99** (1977), 447–485.

7. H. Charalambous and E. G. Evans, *Problems on betti numbers of finite length modules.*

8. A. Kustin, *The minimal free resolutions of the Huneke-Ulrich deviation two Gorenstein ideals*, J. Algebra **100** (1986), 265–304.

9. A. Kustin, *Gorenstein algebras of codimension four and characteristic two*, Comm. Alg. **15** (1987), 2417–2429.

10. A. Kustin, *Classification of the Tor-algebras of codimension four almost complete intersections*, preprint, 1991.

11. A. Kustin, *The minimal resolution of a codimension four almost complete intersection is a DG-algebra*, preprint, 1991.

12. A. Kustin and M. Miller, *Algebra structures on minimal resolutions of Gorenstein rings of embedding codimension four*, Math. Z. **173** (1980), 171–184.

13. A. Kustin and M. Miller, *Algebra structures on minimal resolutions of Gorenstein rings*, Commutative Algebra: Analytic Methods, Lecture Notes in Pure and Applied Math. **68**, ed. R. Draper, Marcel Dekker, 1982.

14. A. Kustin and M. Miller, *Multiplicative structure on resolutions of algebras defined by Herzog ideals*, J. London Math. Soc. (2) **28** (1983), 247–260.

15. A. Kustin and M. Miller, *Classification of the Tor-algebras of codimension four Gorenstein local rings*, Math. Z. **190** (1985), 341–355.

16. S. Palmer, *Multiplicative structure on resolutions of grade four almost complete intersections*, Ph.D. dissertation, University of South Carolina, 1990.

17. C. Peskine and L. Szpiro, *Liaison des variétés algébriques I*, Invent. Math. **26** (1974), 271–302.

18. H. Srinivasan, *Algebra structures on some canonical resolutions*, J. Algebra **122** (1989), 150–187.

19. H. Srinivasan, *Minimal algebra resolutions for cyclic modules defined by Huneke-Ulrich ideals*, J. Algebra **137** (1991), 433–472.

20. H. Srinivasan, *The non-existence of minimal algebra resolution despite the vanishing of Avramov obstructions*, preprint, 1989.

21. B. Ulrich, *Sums of linked ideals*, Trans. Amer. Math. Soc. **318** (1990), 1–42.

22. J. Weyman, *On the structure of free resolutions of length 3*, J. Algebra **126** (1989), 1–33.

Wonderful rings and awesome modules

George R. Kempf / The Johns Hopkins University

Let $A = k \oplus A_1 \oplus \ldots$ be a connected graded commutative ring over a field k. Let M be a finitely generated graded A-module. The M is awesome if $\mathrm{Tor}_i^A(M, k)$ is purely of degree i for all i if M has a free resolution

$$\ldots \xrightarrow{\varphi_1} A^{\oplus n_1} \xrightarrow{\varphi_0} A^{\oplus n_0} \longrightarrow M \longrightarrow 0$$

where all the φ_i are matrices of elements of A_1.

Trivially we have the relation

$$\varphi_A(t) \left(\sum_{i \geq 0}^{\infty} (-1)^i n_i t^i \right) = \varphi_M(t)$$

where φ_* is the Hilbert function of $*$.

The ring A is wonderful if the module k is awesome. The name Koszul seems to be popular with algebraic geometers.

I ask the general problem of giving good examples of these special structures especially those occuring in algebraic geometry. For some other details I refer to the paper [1]. I will give here some examples. See [1] for references.

(Fröberg) if A is generated by elements of degree one $X_1 \ldots X_r$ modulo quadratic monomials, then A is wonderful. This result implies that the usual rings of homogeneous spaces are wonderful; e.g. the Plüker coordinate ring of a Grassmannian. Besides these special rings there are always many wonderful rings in algebra geometry by

(Backelin) if A is generated by a finite number of elements of A_1, then $A^{(d)} = \bigoplus_{i \geq 0}^{\infty} A_{di}$ is wonderful if $d >> 0$. Therefore high enough Veronese embedding of projective varieties have wonderful coordinate rings.

(Kempf) if L is an ample invertible sheaf on an abelian variety X, the ring $\bigoplus_{n \geq 0} \Gamma(X, L^{\oplus n})$ is wonderful if L is an at least 4 power of some N and if M is (algebraically equivalent to) an at least third power of N, the module $\bigoplus_{n \geq 0} \Gamma(X, M \otimes N^{\otimes n})$ is awesome [3].

(Butler unpublished) let C be a complete smooth curve of genus g of characteristic zero. If W is a invertible on C of slope $\geq 2g + 2$, then

$$\bigoplus_{n \geq 0} \Gamma(X, \mathrm{Sym}^n W)$$

is wonderful. Also if U is another invertible sheaf on d of slope $\geq 2g$, then

$$\bigoplus_{n \geq 0} \Gamma(X, U \otimes \mathrm{Sym}^n W)$$

is awesome.

(Kempf [2]) Let $p_1, \ldots, p_r$ be points in $I\!P^n$ where $r \leq 2n$ which are in general linear position. Then the projective coordinate ring of these points is wonderful.

If C as above has no g_2^1, g_3^1 or g_5^1, then the canonical ring $\bigoplus_{n \geq 0} \Gamma(X, \Omega_C^{\otimes n})$. (This is to appear),

It seems natural to study projective coordinate rings of surfaces.

Motivated by the examples of awesome modules over wonderful ring. I asked the following problem. If A is wonderful then for fixed M then define the graded module $M_{[r]}$ by

$$M_{[r],i} = \begin{cases} M_{r+i} & \text{if } i \geq 0 \\ 0 & \text{otherwise.} \end{cases}$$

Then the problem is to show that $M_{[r]}$ is awesome if $r \gg 0$. This problem was solved by Avramov and Eisenbud at the conference.

References

1. G. Kempf, Some wonderful rings in algebraic geometry, Journal of Algebra, to appear.
2. _____, Syzygies of points in projective space, to appear in Journal of Algebra.
3. _____, Projective coordinate rings of abelian varieties, Algebraic Analysis, geometry and Number Theory, Edited J. I. Igusa, The Johns Hopkins Press 1989, pp. 225-236.

Green's Conjecture

Green's Conjecture:

An Orientation for Algebraists

by

David Eisenbud[1]

Department of Mathematics, Brandeis University
Waltham MA 02254
eisenbud@brandeis.bitnet

These notes parallel the introduction to Mark Green's conjecture on the free resolutions of canonical curves (Green [1984] and Green-Lazarsfeld [1985b]) that I presented at the Sundance 90 conference. They have a two-fold purpose: to introduce commutative algebraists with a modest background in algebraic geometry to a formulation of Green's conjecture that is more algebraic than the usual one; and to survey some of the approaches to Green's conjecture that have been tried.

The first section leads up to an algebraic conjecture (somewhat wild) generalizing Green's conjecture. The second section tries to explain the attractiveness of canonical rings of curves (for algebraists; geometers already know this!) and explains the connection between the algebraic conjecture of section 1 and the usual version of Green's conjecture. The third section surveys some promising approaches to Green's conjecture.

1. Ideals generated by quadrics and 2-linear resolutions

Notation

Let $S = k[x_0, \ldots ,x_r]$, and let $R = S/I$ be a homogeneous factor ring of S. If we assume that I contains no linear forms, and the projective dimension of S/I is m, then we may represent the minimal free resolution $\mathcal{F}$ of S/I by a **betti diagram** (as in the "betti" command of the program Macaulay of Bayer and Stillman) of

---

1. Thanks to Brigham Young University, which very generously supported the conference at which these notes originated. I am also grateful to the NSF for partial support during the preparation of this document.

the form

		step of resolution				
		0	1	2	3	...	m
	0	1	–	–	–	...
(degree of syzygy) –	1	–	a_1	a_2	a_3	...	a_m
(step of resolution)	2	–	b_1	b_2	b_3	...	b_m
	3	–	....				c_m

meaning that $\mathcal{F}$ can be written as

$$0 \leftarrow S/I \leftarrow S \leftarrow S(-2)^{a_1} \oplus S(-3)^{b_1} \oplus ... \leftarrow S(-3)^{a_2} \oplus S(-4)^{b_2} \oplus ...$$

$$\leftarrow .... \leftarrow S(-(m+1))^{a_m} \oplus S(-(m+2))^{b_m} \oplus S(-(m+3))^{c_m} \oplus ... \leftarrow 0.$$

Here the dashes in the betti diagram represent zeros. We define the 2-linear strand of $\mathcal{F}$ to be the subcomplex

$$0 \leftarrow S/I \leftarrow S \leftarrow S(-2)^{a_1} \leftarrow S(-3)^{a_2} \leftarrow .... \leftarrow S(-(m+1))^{a_m} \leftarrow 0.$$

Note that if some a_i is 0 then, since $\mathcal{F}$ is minimal, all a_j for $j \geq i$ are 0 as well. Thus we define the **length** of the 2-linear strand to be the largest number n such that $a_n \neq 0$. For simplicity we will call this this n the **2-linear projective dimension** and write

2LP (S/I) = n.

The Theme

In general, not too much is known about the linear strand and its length (but see Eisenbud-Koh [1991?a,b] for some conjectures in a slightly different context.) However, the theme of Green's conjecture is that long linear strands in free resolutions of nice ideals have a fairly simple and uniform origin. To understand it consider first a very easy and well-known result:

Lemma 1 . 1: If $I \subset J$ are ideals containing no linear forms, then the 2-linear strand of the minimal free resolution $\mathcal{F}$ of S/I is a summand of the 2-linear strand of the resolution $\mathcal{G}$ of S/J. In particular, the length of the 2-linear strand of $\mathcal{G}$ is $\geq$ that of the 2-linear strand of $\mathcal{F}$.

Proof: Let $\varphi \colon \mathcal{F} \to \mathcal{G}$ be a comparison map covering the canonical projection $S/I \to S/J$. We claim that the map on the ith step of the resolution, $\varphi_i \colon \mathcal{F}_i \to \mathcal{G}_i$, induces a split monomorphism on the 2-linear part of $\mathcal{F}$.

At the 0th step $\mathcal{F}_0 = \mathcal{G}_0 = S$, and the result is clear. Since J contains no linear forms, the quadratic minimal generators in the ideal I are among the minimal generators of J; this is exactly equivalent to the desired statement. An easy induction completes the proof. //

Thus one kind of acceptable "explanation" for the length (or size...) of the 2-linear part of the resolution of S/I would be that I contains an ideal of some standard form, which is known to have a long (or large...) 2-linear part. This is the form which Green's conjecture takes. Before stating the conjecture, we will describe the ideals of "standard form" that arise:

Examples

If I is the ideal generated by the 2×2 minors of a generic $p \times q$ matrix, then the 2-linear strand is known (Lascoux [1978]) to have length $\geq p+q-3$, with equality in characteristic 0 (and most likely all the time.) This turns out to be true of ideals of 2×2 minors of considerably more general matrices: following Eisenbud [1988] we define a matrix $L = (\lambda_{ij})$ of linear forms to be **1-generic** if no entry λ_{ij} is 0, and none can be made 0 by row and column transformations. We have

Theorem 1 . 2: If L is a 1-generic matrix of linear forms over a polynomial ring S, and I is the ideal generated by the 2×2 minors of L, then the minimal free resolution of S/I has a 2-linear strand of length $\geq p+q-3$.

This result was proved under various extraneous hypotheses by Green and Lazarsfeld (Green [1984;Appendix]); in a more algebraic setting by myself (unpublished); and then, in a simpler way, by Koh and Stillman [1989]. The most familiar case of Theorem 1 . 2 is that in which one of p, q, say p, is 2: Assuming that if the 2×q matrix of linear forms L is 1-generic, it is not hard to show that the ideal of minors of L is of generic codimension (= q-1), so the minimal free resolution $\mathcal{F}$ is the the Eagon-Northcott complex, which consists entirely of a linear strand of length exactly p+q-3 = q-1. For arbitrary p,q and a generic matrix L, the computation of Lascoux shows that the 2-linear strand has length exactly p+q-3; the condition of 1-genericity is sufficient (but is not necessary; it would be nice to know some necessary conditions) to preserve at least one of the required syzygies.

An algebraic conjecture:

The boldest possible conjecture would now be to say that a sort of converse to Theorem 1 should be true: that is, any ideal I such that 2LP(S/I) = n should contain an ideal of 2×2 minors of a 1-generic p×q matrix with p+q-3 = n. Alas, this is <u>false</u>.

First of all, it is possible to start with a 1-generic matrix and replace some of its entries with 0's without spoiling the length of the 2-linear part of the resolution of its ideal of 2×2 minors, so that some matrices which are not 1-generic might perform the same function. However, if we assume that I is a prime ideal not containing any linear forms, then of course I cannot contain a nonzero determinant of a 2×2 matrix of linear forms with one entry = 0. Thus I could not contain a nontrivial ideal of 2×2 minors of any matrix of linear forms other than a 1-generic matrix (or a 1-generic matrix expanded with rows and columns of zeros.) So if we assume that I is prime, this objection becomes void.

Second, there are ideals whose 2-linear part is nontrivial but which contain no determinants at all!. For note that any determinant of a 2×2 matrix of linear forms is a quadric of rank ≤ 4 (the rank must be 3 or 4 if the matrix is 1-generic.) Thus the ideal I generated by a single rank s quadric, with s≥5, has a 2-linear part of length 1 but certainly contains no 2×2 minors of interest. The same holds for some quadrics of rank ≤ 4 if the field is not algebraically closed: for example $x^2+y^2+z^2+w^2$ is not a determinant of a matrix of linear forms over the real numbers.

But what about prime ideals generated by quadrics of rank ≤ 4 (or more generally, primes whose degree 2 parts are spanned by quadrics of rank ≤ 4) over an algebraically closed field? There are still counterexamples, due to Schreyer [1986, 199?], in characteristic 2 (analogous cases, in characteristics ≠ 2, are known not to be counterexamples.) However, one may be optimistic and feel that the theory of ideals with lots of quadratic generators might well be a little different in characteristic 2. (Of course the pessimistic will feel instead that we are being warned.) If we suppose that char k = 0, or even that it is ≠ 2, I know no further counterexamples, so I rashly make the

Conjecture 1 . 3: Let k be an algebraically closed field of characteristic ≠ 2, and let $I \subset S = k[x_0, \ldots ,x_r]$ be a prime ideal, containing no linear form, whose quadratic part is spanned by quadrics of rank ≤ 4. If 2LP(S/I) = n, then I contains an ideal of 2×2 minors of a 1-generic p×q matrix with p+q-3 = n.

Green's conjecture, from the algebraic point of view, is just the special case of this where we assume <u>in addition</u>:

    a) S/I is normal (= integrally closed)
    b) dim S/I = 2
    c) S/I is Gorenstein
    d) degree S/I = 2r

(conditions c,d could be replaced by the equivalent condtions that S/I is a Cohen-Macaulay domain such that modulo a maximal regular sequence of linear forms it has Hilbert function 1, γ, γ, 1 for some integer γ, which is r-1 in the notation above.)

There is really no argument connecting any of these four conditions to the conclusion of the conjecture; but there are some geometric techniques (described below) which have lead to the verification of Green's conjecture under these extra hypotheses in many special cases (for example in all cases with r ≤ 7; see Schreyer [1986].)

2. Canonical rings of curves

In the previous section I claimed that the special hypotheses a) – d) might well be irrelevant to the conjecture. In this section I want to explain why geometers nevertheless find a)-d) so entrancing, and describe a geometric reformulation under these conditions.

What is the simplest interesting kind of variety? A curve, of course! (Well, almost of course: rational varieties also have exercised some claims, especially in certain periods, because they are so easy to specify. For example a rational surface can be specified as a plane with some marked points (to blow up) and curves (to blow down.) But for our purposes here, the anwer is certainly "a curve.") It's natural to consider first only nonsingular projective curves; and one finds that the set of isomorphism classes of these curves breaks up by genus into well-behaved algebraic varieties, the "moduli spaces."

Of course any algebraist would rather have a ring than a variety. Do these curves give rings? Not immediately. The simplest and most attractive way for a ring to come from a variety is as the homogeneous coordinate ring of that variety **in some projective embedding.** Thus to get a ring, one wants not only a curve but an embedded curve. The space of embeddings of a curve is not too bad, but there is a simpler way out of this dilemma than studying all embeddings: Leaving aside the so-called **hyperelliptic** curves, which are in any case quite well understood, every curve comes with a uniquely distinguished embedding in projective space, called, for obvious reasons, the **canonical embedding** -- that is, there really is a distinguished **canonical ring**, the homogeneous coordinate ring of the the canonically embedded curve, corresponding to each abstract (non-hyperelliptic) nonsingular curve[2]. With more sophistication the hyperelliptic curves can be

---

2. Other embeddings are interesting too. See for example Eisenbud [transcanonical], and Martens-Schreyer [1986] for some interesting cases, and Green-Lazarsfeld [1985b] for some general conjectures.

included here too, but we will not worry about this point.

Moreover, the canonical rings of curves turn out to be quite simple: they are (precisely) the graded domains which are algebras over a field k and satisfy properties a)-d) from the last section. The number γ introduced there is g-2, where g is the genus of the curve. The fact that the canonical rings of curves are Cohen-Macaulay is called Noether's theorem (see Arbarello et al [1985]). The Gorenstein property follows easily from this using duality theory. That the quadratic part of the ideals is spanned by quadrics of rank 4 was conjectured by Andreotti and Mayer and proved in general by Green [1984] (see also Smith and Varley [1989].)

Let us now take a look at the free resolution of the canonical ring of a curve of genus g (as a module over the homogenous coordinate ring of the projective space in which the curve is canonically embedded: if the curve has genus g, then this is S = $k[x_0, \dots ,x_r]$, with r = g-1.) From the Gorenstein property and property d) it follows that the resolution has betti diagram of the form:

	0	1	2	...	g-4	g-3	g-2
0	1	-	-	...	-	-	-
1	-	a_1	a_2	...	a_{g-4}	a_{g-3}	-
2	-	a_{g-3}	a_{g-4}	...	a_2	a_1	-
3	-	-	-	...	-	-	1

Note the symmetry between the ith row and the $(3-i)^{th}$ row, and that b_i (in our previous diagram) is now given as a_{g-2-i}. Also from the Hilbert function in d) one computes:

$$a_i - a_{g-1-i} = i\binom{g-2}{i+1} - (g-1-i)\binom{g-2}{i-2} \text{ for } i = 1 \dots g-2,$$

so that in fact all the betti numbers are known if we know

(∗) $a_{\lfloor g/2 \rfloor}, \dots , a_{g-3}.$

As we have already mentioned, the vanishing of a_i implies that of a_{i+1}. Similarly if a_i is 1 a_{i+1} vanishes (probably $a_{i+1} \neq 0$ actually implies that a_i is rather large.) For small g and in certain other cases we know a lot about the sequence (*). However, in general, we know little about which sequences of numbers (*) can occur. The same remarks are valid for the free resolution of any homogeneous R = S/I satisfying hypotheses c) and d) from section 1 -- the assumptions that R is 2-dimensional, normal or even a domain, are irrelevant. Of course the possibilities for the sequence (*) may well expand if we weaken the hypotheses in this way. But also in the general case, we do not know which sequences (*) occur. It is not even known whether the sequence can be 0, ... , 0! In fact a leading special case of Green's conjecture (the so-called "generic case") is just this:

Generic Green's Conjecture 2 . 1: There exists a homogeneous Gorenstein ring (respectively a homogeneous Gorenstein normal domain) as above with

$$a_{\lfloor g/2 \rfloor} = 0$$

(and thus a_i = 0 for i ≥ [g/2].)

We will call the two versions of this conjecture, with and without the "normal domain" condition, the strong and weak forms, respectively. The strong form is known to hold for g ≤ 17, by virtue of computer work (at least for some characteristic, and a fortiori for almost all characteristics, including characteristic 0) using the approach of Weyman detailed below. It is also known that each of the cases (g odd) and (g even) implies the other (work of Ein, Bayer-Stillman and Weyman...).

To massage Green's conjecture into its usual statement, and to relate it to Conjecture 1 . 3, we quote a (rather easy) result derived from our [1988, pp. 549-552]:

Theorem 2 . 2: Let $X \subset \mathbb{P}^r$ be a reduced irreducible linearly normal nondegenerate curve. There is a 1-generic p×q matrix of linear forms L whose 2×2 minors vanish on X iff the hyperplane bundle $\mathcal{O}_X(1)$ can be factored as $\mathcal{L}_1 \otimes \mathcal{L}_2$ with $h^0\mathcal{L}_1 \geq p$ and $h^0\mathcal{L}_2 \geq q$. //

Here the conditions "linearly normal nondegenerate" mean that the natural map $H^0(\mathcal{O}_{\mathbb{P}^r}(1)) \to H^0(\mathcal{O}_X(1))$ is an isomorphism.

Recall that the Clifford index of a line bundle $\mathcal{L}$ on a curve C is by definition the number

Cliff $\mathcal{L} := g+1 - (h^0(\mathcal{L}) + h^0(\omega \otimes \mathcal{L}^{-1})) = \deg \mathcal{L} - 2(h^0(\mathcal{L})-1)$

and that the Clifford index of C (in case the genus of C is ≥ 3) is the minimum of all Clifford indices of bundles $\mathcal{L}$ with $h^0\mathcal{L} \geq 2$ and $h^0\omega \otimes \mathcal{L}^{-1} \geq 2$. Putting this together, we get

Corollary 2 . 3: Let C be a curve of Clifford index c canonically embedded in $\mathbb{P}^{g-1}$. If c' is the maximum number such that there is a 1-generic p×q matrix of linear forms L whose 2×2 minors vanish on X and p+q-3 = c', then c' = g-2-c. //

Thus Conjecture 1 . 3 becomes, in the case of canonical curves:

Conjecture 2 . 4 (Green):   The length of the 2-linear part of the resolution $\mathcal{F}$ of the canonical ring S/I of a curve of genus g and Clifford index c is

2LP (S/I) = g-2-c.

Equivalently, with notation as in section 1 above, we have

b_1 = ... = b_{c-1} = 0, but $b_c \neq 0$.

In Green's terminology, this says that a curve of Clifford index c satisfies condition N_{c-1} but not N_c. Note that the fact that the length of the linear part is at least g-1-c (equivalently $b_c \neq 0$) comes "for free" from Theorems 1 . 2 and
≡≡ Cited Item Has Been Deleted ≡≡; this "easy half" was first proved by Green and Lazarsfeld in the appendix to Green's [1984].

The generic curve of genus g is known to have Clifford index $\lfloor(g-1)/2\rfloor$ -- in the sense that this is the value of the Clifford index taken on by the curves in an open dense set of the moduli space. Thus the generic form of Green's conjecture becomes:

Generic Green's Conjecture (geometric version) 2 . 5: The free resolution of the canonical ring of a generic curve of genus g has

$$a_{\lfloor g/2 \rfloor}, \, ... \, , a_{g-3} = 0, \, 0, \, ... \, , 0.$$

Since the condition in the conclusion of the conjecture is a Zariski open condition in families of curves, it would be the same to assert that there exists a smooth curve whose canonical ring satisfies this conclusion -- or even a locally Gorenstein, smoothable, one-dimensional, canonically embedded scheme with this property.

3. Special cases and general approaches

In this section we will list the known approaches to Green's conjecture. In order to give something absolutely explicit, we summarize an approach due to Weyman giving generators and relations for certain graded modules of finite length over polynomial rings whose hilbert functions determine the desired betti numbers. We apologize in advance to the expert whose favorite bit we skipped.

I. Degenerations and the strong generic conjecture

This section describes approaches to the "generic" Green's conjecture (1.2 or 1.5 above) using the technique of degenerations.

These approaches to the (strong form of the) generic Green's conjecture lead quite quickly to explicit ideals of canonical curves, one for each genus, which experimentally have the kind of resolution predicted by Green's conjecture for the generic curve. If one could prove that one of these ideals really has the desired resolution, then one would have a proof of the generic Green's conjecture, since each of these examples represents a "degeneration"

of a smooth canonical curve of genus g. There is even one version of
this, coming from the Ribbon approach of Bayer-Eisenbud, that
would lead to a proof that Green's conjecture holds for a curve of
each Clifford index.

There is an interesting opportunity for an algebraist here. All
that is involved is to find the free resolution of a particular ideal. So
far at least, geometric ideas have not been of much use.

Though the problems involved are of the most concrete and
specific kind (specific ideals whose minimal resolutions must be
calculated -- sometimes with explicitly known non-minimal
resolutions, so that the computation comes down to deciding the
rank of an explicitly known matrix with integer entries!), people
have not written much up.

A. Cuspidal rational curves and the tangent developable surface. (Buchweitz, Schreyer, Bayer-Stillman, Weyman, Green, Kempf...)

The following construction was noted by Buchweitz and Schreyer
some time around 1983: One way to get a canonical curve of genus
g is to take a rational curve with g cusps in $\mathbb{P}^{g-1}$. Such curves turn
out to be the hyperplane sections of an arithmetically Cohen-
Macaulay surface (actually a degenerate K3 surface, if there are
any geometers listening) obtained as the tangent developable
surface ($\equiv$ the union of the tangent lines) to the rational normal
curve in $\mathbb{P}^g$. This is the rational surface with affine parametrization
$\mathbb{A}^2 \to \mathbb{A}^g$ given by

$$(x,y) \mapsto (x,x^2,...,x^g) + y(1,2x,3x^2,...,gx^{g-1})$$

$$= (x+1, x^2+2xy, \ldots).$$

Since these curves are smoothable, a proof of the generic version of
Green's conjecture would follow if we could check the conjecture on
the resolution of any one of these curves, or, equivalently, on the
resolution of the tangent developable surface itself.

We will show below that if we take the rational normal curve to have equations given by the minors $\Delta_{a,b}$ of the matrix

Let $\Delta_{a,b}$ $(0 \leq a, b \leq g-1)$ be the minor involving the columns a and b of the matrix

$$\begin{bmatrix} x_0 & x_1 & \cdots & x_{g-1} \\ x_1 & x_2 & \cdots & x_g \end{bmatrix}$$

with the usual convention that $\Delta_{a,b} = -\Delta_{b,a}$. If we take the rational normal curve to be the curve with equations

$$\Delta_{a,b} = 0 \qquad\qquad (0 \leq a, b \leq g-1),$$

then we will show below that the quadratic equations of the tangent developable surface are

$$\Gamma_{a,b} := \Delta_{a+2,b} - 2\Delta_{a+1,b+1} + \Delta_{a,b+2} = 0 \quad (0 \leq a < b \leq g-3).$$

(For $g \geq 6$ the quadratic equations generate the ideal of the tangent developable surface, but in any case only the quadratic equations are involved in checking Green's conjecture.) Thus:

To prove the (strong) generic version of Green's conjecture, and indeed to prove the conjecture for generic smooth curves, it is enough to show that the 2-linear part of the minimal resolution of the ideal J generated by the linear combinations

$$\Delta_{a,b+2} - 2\Delta_{a+1,b+1} + \Delta_{a+2,b} \qquad 0 \leq a < b \leq g-3$$

of the minors of the matrix M has length $\leq$ (or, equivalently, $=$) $\lfloor g/2 \rfloor - 1$.

We will now derive these equations for the tangent developable surface. We will then exhibit an explicit non-minimal free resolution for the ideal J in a form discovered by Jerzy Weyman. All this will be done in terms of multilinear algebra and the representation theory of SL_2; the reader without a taste for such constructions may wish to skip to the description of "Weyman's Modules", below:

We begin with an invariant description of the ideal of the rational normal curve. We think of $\mathbb{P}^1$ as $\mathbb{P}(V)$, the space of 1-quotients of a 2-dimensional vector space V, and write Aut $\mathbb{P}^1$ = SL(V) = SL_2 . The ambient space $\mathbb{P}^g$ of the rational normal curve is then $\mathbb{P}(S_gV)$, where S_g denotes the gth symmetric power functor, and we think of S_gV as the forms of degree g on $\mathbb{P}^1$. The space of all quadratic forms on this $\mathbb{P}^g$ is $S_2(S_gV)$. The restriction map of these to the rational normal curve (= $\mathbb{P}^1$) is then the natural map

$$S_2S_gV \rightarrow S_{2g}V,$$

and the quadratic part of the ideal of the rational normal curve is the kernel of this map. It is convenient to denote a basis of V by $\{1,x\}$, and more generally a basis for S_gV by $\{1, x, \ldots, x^g\}$, in which case the map $S_2S_gV \rightarrow S_{2g}V$ above may be written

$$x^a \cdot x^b \mapsto x^{a+b}.$$

In characteristic 0 the kernel is easy to describe and analyze, as follows: We have

$$S_2S_gV = S_{2g}V \oplus S_{2g-4}V \oplus S_{2g-8}V \oplus \ldots .$$

The quadratic part of the ideal of the rational normal curve itself consists of all but the first of these summands. It is thus

$$S_2S_{g-2}V = S_{2g-4}V \oplus S_{2g-8}V \oplus \ldots ,$$

and (as we shall show) the quadratic part of the ideal of the tangent developable surface consists of all but the first two. It is thus

$$S_2S_{g-4}V = S_{2g-8}V \oplus S_{2g-12}V \oplus \ldots .$$

In both cases the inclusion maps are given by "inner multiplication" by

$$\alpha := x^2 \cdot 1 - x \cdot x \in S_2S_2V,$$

where by inner multiplication we mean the equivariant pairing

$$S_2S_aV \otimes S_2S_bV \to S_2S_{a+b}V$$

given by

$$x^u{\cdot}x^v \otimes x^s{\cdot}x^t \mapsto x^{u+s}{\cdot}x^{v+t} + x^{u+t}{\cdot}x^{v+s}.$$

Unfortunately, this description does not work in arbitrary characteristic, so we adopt another, only a little less simple which does (I am grateful to Jerzy Weyman for discussions of this matter):

We can identify the space Q of quadratic forms in the ideal of the rational normal curve with the representation $\wedge^2 S_{g-1}V$ (which is the same as $S_2S_{g-2}V$ in characteristic 0). To get the inclusion map, note that the kernel of $S_2S_gV \to S_{2g}V$ is spanned by the elements

$$\Delta_{a,b} = x^ax^{b+1} - x^{a+1}x^b \in S_2S_gV \qquad \text{for } 0 \le a < b \le g{-}1,$$

which are the images of the obvious basis vectors

$$x^a{\wedge}x^b \in \wedge^2 S_{g-1}V$$

(this is half the image of the product of $(x \otimes 1 - 1 \otimes x)$ and $x^a \otimes x^b - x^b \otimes x^a$ in $S_gV \otimes S_gV$, under the map to S_2S_gV; the map is thus equivariant.)

We claim next that the quadratic part of the ideal of the tangent developable surface is given by the representation

$$\wedge^2 S_{g-3}V \subset \wedge^2 S_{g-1}V.$$

As in the characteristic 0 description, this map can be described as an "inner multiplication with the element α", where this time inner multiplication refers to the pairing

$$\wedge^2 S_aV \otimes S_2S_bV \to \wedge^2 S_{a+b}V$$

given by

$$x^u \wedge x^v \otimes x^s \cdot x^t \mapsto x^{u+s} \wedge x^{v+t} + x^{u+t} \wedge x^{v+s}.$$

The image of $x^a \wedge x^b$ in $S_2 S_g V$ is now easily computed to be

$$\Gamma_{a,b} = \Delta_{a+2,b} - 2\Delta_{a+1,b+1} + \Delta_{a,b+2} \quad \text{for } 0 \le a < b \le g-3$$

We now claim that the quadratic part of the ideal of the tangent developable surface is generated by these. It is useful to fall back on the characteristic 0 computation: since $\alpha * S_2 S_{g-2} V$ generates a maximal invariant subspace of $S_2 S_g V$, the same will be true of
$S_2 S_{g-4}$ in $S_2 S_{g-2}$. A direct computation shows that the image of $S_2 S_{g-4}$ in $S_2 S_g$ is then spanned by the elements

$$\Gamma_{a,b+1} - \Gamma_{a+1,b} \quad \text{for } 0 \le a \le b \le g-4$$

-- this is the image of $x^a \cdot x^b$. But over $\mathbb{Q}$ these span the same space as the elements $\Gamma_{a,b}$ themselves. Thus over $\mathbb{Q}$ the $\Gamma_{a,b}$ span a maximal invariant subspace of the ideal of the rational normal curve, and it suffices to show that they vanish on the tangent developable surface. By invariance it is enough to show that these forms vanish on a single tangent line to the rational normal curve. For example, the tangent line at the point "x=0", that is, $(1,0, \ldots ,0)$, is the line consisting of all elements $(s, t, 0, \ldots ,0)$. This vanishing occurs because the only minor $\Delta_{u,v}$ not vanishing on this line is $\Delta_{0,1}$, and this one never occurs in the expression for $\Gamma_{a,b}$.

We have now shown that the representation $\wedge^2 S_{g-3} V \subset S_2 S_g V$ defines the quadratic part of the ideal of the tangent developable surface in characteristic 0. To complete the proof in all characteristics it suffices to show that, if we take V for a moment to be $\mathbb{Z}^2$, the quotient $S_2 S_g V / (\Gamma_{a,b} \mid 0 \le a < b \le g-3)$ is torsion free, which amounts to finding a minors of the appropriate size in the inclusion matrix which are relatively prime. This is a straightforward computation.

Because the ideal J of the tangent developable surface is SL_2-

invariant, it is natural to try to write down a resolution in terms of representation theory; for example, the Koszul cohomology groups might be written in terms of representations of SL_2. Several people seem to have tried this without success; an invariant but nonminimal resolutions are known, but not how to make any of them minimal! One such construction is detailed in the paper of Bayer and Stillman elsewhere in these proceedings.

Here is such a construction of a nonminimal resolution, from Weyman's private notes (about 1986). It turns out that this is a special case of the construction pursued in Weyman's paper [1989], (where the ideal J of the tangent developable is called J_{g-1}) to which the reader may go for more details.

Let R be the polynomial ring $k[S_gV]$, and let J be the ideal of the tangent developable surface, as above. Weyman's idea is based on a computation of the free resolution of the normalization A of R/J, realized by pushing forward the structure sheaf from the desingularization of the tangent developable surface. He finds:

1) A decomposes as SL(V)-module as

$$A = \oplus_{d \geq 0} S_{(g-1)d}V \otimes S_d V.$$

2) The cokernel C := A/(R/J) decomposes as

$$C = \oplus_{d \geq 1} S_{dg-2}V \otimes \wedge^2 V.$$

3) The minimal free resolution of A over R is

$$0 \leftarrow A \leftarrow S_0V(0) \oplus \wedge^2 V \otimes S_{g-2}V(-1) \leftarrow \wedge^2 V \otimes \wedge^2(S_{g-2}V) \otimes S_2V(-2)$$

$$\ldots \leftarrow \wedge^2 V \otimes \wedge^{i+1}(S_{g-2}V) \otimes S_{2i}V(-i-1) \leftarrow \ldots$$

(The terms are to be thought of as the free R-modules obtained by tensoring the given representation, over k, with R.) Note that SL(V)-equivariantly the term $\wedge^2 V = k$ could have been suppressed, but has been carried along to make the descriptions of the subsequent maps simpler; it also makes the description GL(V)-

equivariant. The differential between the terms of the form $\wedge^2 V \otimes \wedge^i(S_{g-2}V) \otimes S_{2i-2}V(-i)$ is the composite

$$\wedge^2 V \otimes \wedge^{i+1}(S_{g-2}V) \otimes S_{2i}V(-i-1)$$

$$\Big\downarrow 1 \otimes \Delta \otimes \Delta$$

$$\wedge^2 V \otimes \wedge^i(S_{g-2}V) \otimes S_{g-2}V \otimes S_2 V \otimes S_{2i-2}V(-i-1)$$

$$\Big\downarrow 1 \otimes 1 \otimes \text{mult} \otimes 1$$

$$\wedge^2 V \otimes \wedge^i(S_{g-2}V) \otimes S_g V \otimes S_{2i-2}V(-i-1)$$

$$\Big\downarrow \text{the factor } S_g V \text{ is absorbed into the coefficients}$$

$$\wedge^2 V \otimes \wedge^i(S_{g-2}V) \otimes S_{2i-2}V(-i)$$

where Δ represents the diagonal map on the symmetric or exterior algebra, mult is the natural map $S_{g-2}V \otimes S_2 V \to S_g V$, and the last map is the one which multiplies the factor $S_g V$ with the "coefficients", in $k[S_g V]$, and raises the degree by 1.

The remaining piece of the differential,

$$S_0 V(0) \leftarrow \wedge^2 V \otimes \wedge^2(S_{g-2}V) \otimes S_2 V(-2)$$

comes from a higher map in the spectral sequence which gives rise to this resolution. Thus we know its definition only on the kernel of the map

$$\wedge^2 V \otimes \wedge^2(S_{g-2}V) \otimes S_2 V(-2) \to \wedge^2 V \otimes S_{g-2}V(-1)$$

which is the part of the differential we have already defined. Since it plays no role in the computations necessary to check Green's conjecture, we will worry about it no further.

4)   The minimal free resolution over R of C is

$$0 \leftarrow C \leftarrow \wedge^2 V \otimes S_{g-2} V(-1) \leftarrow (\wedge^2 V)^{\bullet 2} \otimes S_{g-3} V \otimes S_{g-1} V(-2)$$

$$\ldots \leftarrow (\wedge^2 V)^{\bullet i+1} \otimes S_{g-i-2} V \otimes \wedge^i (S_{g-1} V)(-i-1) \leftarrow \ldots$$

The differential is the composite

$$(\wedge^2 V)^{\bullet i+1} \otimes S_{g-i-2} V \otimes \wedge^i (S_{g-1} V)(-i-1)$$

$$\Big\downarrow \quad [(\wedge^2 V)^{\bullet i+1} = (\wedge^2 V)^{\bullet i} \otimes \wedge^2 V] \otimes 1 \otimes \Delta$$

$$(\wedge^2 V)^{\bullet i} \otimes \wedge^2 V \otimes S_{g-i-2} V \otimes \wedge^{i-1}(S_{g-1} V) \otimes S_{g-1} V(-i-1)$$

$$\Big\downarrow \quad 1 \otimes \eta \otimes 1 \otimes 1$$

$$(\wedge^2 V)^{\bullet i} \otimes V \otimes S_{g-i-1} V \otimes \wedge^{i-1}(S_{g-1} V) \otimes S_{g-1} V(-i-1)$$

$$\Big\downarrow \quad \text{multiply } V \text{ and } S_{g-1} \text{ to } S_g V \text{ and absorb}$$

$$(\wedge^2 V)^{\bullet i} \otimes S_{g-i-1} V \otimes \wedge^{i-1}(S_{g-1} V)(-i),$$

where $\eta: \wedge^2 V \otimes S_{g-i-2} V \to V \otimes S_{g-i-1} V$ is the composite

$$\wedge^2 V \otimes S_{g-i-2} V$$

$$\Big\downarrow \quad \Delta \otimes 1$$

$$V \otimes V \otimes S_{g-i-2} V$$

$$\Big\downarrow \quad 1 \otimes 1 \otimes \text{mult}$$

$$V \otimes S_{g-i-1} V$$

as in the Koszul complex.

5) The resolution of R/J itself is now obtained from a mapping cylinder for a comparison map Θ from the resolution of A to the resolution of C lifting the natural epimorphism A → C. Let us write

$A_i = \wedge^{i+1}(S_{g-2}V) \otimes S_{2i}V$ (for i > 0; A_0 has another term, = S_0V)

$C_i = S_{g-i-2}V \otimes \wedge^i(S_{g-1}V)$

for the ith term in the resolutions of A and C respectively, where for simplicity we have suppressed the factors of $\wedge^2V$ and the twist (-i-1). We wish to give an explicit expression for the map Θ on the ith term of the resolution,

$$\Theta_i: A_i \rightarrow C_i.$$

To this end we denote a basis of each S_mV by powers of x, as above. Let $\Delta: S_{2i}V \rightarrow (S_2V)^{\otimes i}$ be the diagonal map, and write

$$\Delta(w) = \Sigma_s \, w_s^1 \otimes ... \otimes w_s^i \,.$$

The map Θ_i is given by the formula

$$\Theta_i(x_{j_1} \wedge ... \wedge x_{j_{i+1}} \otimes w)$$

$$= \Sigma_s \Sigma_{1 \leq a_2 \leq j_2 - j_1, \, ... \, , \, 1 \leq a_{i+1} \leq j_{i+1} - j_i}$$

$$x^{j_1 + a_2 + ... + a_{i+1} - i} \otimes \{ \, x^{j_2 - a_2} \cdot w_s^1 \wedge ... \wedge x^{j_{i+1} - a_{i+1}} \cdot w_s^i \, \}$$

for $0 \leq j_1 < ... < j_{i+1} \leq g-2$ and $w \in S_{2i}V$.

6)   We see that R/J has a nonminimal resolution of the form

$$0 \leftarrow R/J \leftarrow R \leftarrow \ker(\Theta_1)(-2) \oplus \operatorname{coker}(\Theta_2)(-3)$$

$$... \leftarrow \ker(\Theta_i)(-2) \oplus \operatorname{coker}(\Theta_{i+1})(-3) \leftarrow ...$$

$$... \leftarrow R(-g-1) \leftarrow 0$$

Thus to prove the generic Green's conjecture, it is enough to show that the map Θ_i of vector spaces given explicitly above is onto for i $\leq \lfloor g/2 \rfloor$.

B. Weyman's Modules:

Making use of the representation theory in a different way, and working over a field k of characteristic 0, Jerzy Weyman (1988, unpublished) has defined for each i $\geq$ 2 a graded module which I will call W(i), generated in degree 2 and having finite length over the polynomial ring in i+1 variables. It has the following remarkable property: the Hilbert function of W(i) determines the number we called a_{i-1} in section 1 for <u>all</u> the tangent developable surfaces, and thus for the cuspidal rational curves of <u>every</u> genus g. The mixing of the different genera in a single module seems to me quite surprising. I am not aware of any geometric explanation of these modules.

To be specific, suppose we are dealing with a g cuspidal curve, and write [i,j] for the "Newton coefficient" (i+j)!/i!j!. Weyman proves:

$$a_i = (2i+1)[g-i-2,i+1] - (i+2)[g-i-1,i+1] + [g-i,i+1] + \dim_k W(i+1)_{g-i}.$$

The strong form of the generic Green's conjecture can thus be seen to hold in odd genus g = 2h+1 as soon as $W(h+1)_{h+1}$ = 0, and in even genus g = 2h as soon as $W(h+1)_{h-1}$ = 0.

The simple existence of such modules W(i) has some notable consequences: First, since W(h+1) is generated in degree 2, Weyman concludes that the generic Green's conjecture in the even genus case implies the corresponding conjecture in the odd genus case. (Bayer, Stillman, and Ein report that they have proven the reverse implication, so that the even and odd genus cases are actually equivalent.)

Second, since W(i+1) has finite length, Weyman concludes that for sufficiently large g the general curve of genus g has the "correct" i^{th} syzygy (sharper versions of this have been proven by others --

see below.)

The modules $W(i)$ are quite easy to write down in terms of generators and relations: Let V be a 2-dimensional vector space as above, and write S_m for the $m+1$-dimensional vector space which is the mth symmetric power of V, considered as a representation of $SL(V)$. There is a (unique up to scalars) inclusion of representations

$$h: S_{2i-2} \to \wedge^2 S_i \cong S_{2i-2} \oplus S_{2i-6} \oplus S_{2i-10} \oplus \dots .$$

Of course writing $T = k[S_i]$ for tie polynomial ring on the generators of S_i, we may tensor with T and regard h as a map of free T-modules

$$h: T \otimes S_{2i-2} \to T \otimes \wedge^2 S_i.$$

On the other hand, the Koszul complex contains a map of degree 1

$$\kappa: T \otimes \wedge^3 S_i \to T \otimes \wedge^2 S_i.$$

The module $W(i)$ is then generated by the elements of $\wedge^2 S_i$ (regarded as having degree 2) subject to the degree 0 relations given by h and the degree 1 relations given by κ; or in other words,

$$W(i) = \text{coker } (h,\kappa) : T \otimes S_{2i-2} \oplus T \otimes \wedge^3 S_i \to T \otimes \wedge^2 S_i.$$

To make this completely explicit, we give a formula for h. After multiplying each column by a certain integer to avoid denominators (this is a harmless operation in characteristic 0!) we may write h as an integral matrix

$$h = (h_{[p,q], j}) \qquad\qquad (0 \le p < q \le i, \qquad 0 \le j \le 2i-2)$$

defined recursively by

$$h_{[0,1],0} = 1;$$

$$h_{[p,q],j} = (i-p+1)h_{[p-1,q]} + (i-q+1)h_{[p,q-1]}$$

where the terms $h_{[s,t],r}$ with $s < 0$ or $t \leq s$ are to be interpreted as 0.

Using these formulas and the program Macaulay of Bayer-Stillman, a computer has checked the generic Green's conjecture (strong form) for $g \leq 17$ in char 31991 (and thus char 0, and almost every other characteristic.) This is computationally the most effective technique currently known.

Other degenerate canonical curves have also been studied in the hope of getting at the strong generic form of the conjecture. These include:

C. **Nodal rational curves:** Two rational normal curves meeting in $g+1$ points (Bayer, Stillman -- reported elsewhere in these proceedings.)

Bayer and Stillman have tried to compute the Koszul cohomology groups giving the betti numbers of the canonical embedding of a rational cuspidal (or nodal) curve by regarding the canonical series on the cuspidal curve as an incomplete series on the normalization, and computing Koszul cohomology. In this way one obtains very explicit integer matrices, and the generic form of Green's conjecture would be proved by showing that they have certain ranks. This approach will be described in a separate article by them.

D. **Graph curves** (Bayer, Eisenbud, Park [1991])
E. **Ribbons** (Bayer, Eisenbud [199?], Fong [1991])

II. **Large genus compared to Clifford index.**

L. Ein [1987]: an induction formula for the b_j (in the notation of section 1) passing from generic curves of low genus to generic curves of somewhat higher genus.

Schreyer has proved that the conjecutre holds sharply for p-gonal curves as long as p is small compared to the genus (roughly

$g \geq p^2 +$ a linear function of p

is required (this effective version of Weyman's result was found independantly.) Ferola (unfinished thesis) has given a better bound, on the order of $p^2/2$.

Voisin (for g≥ 11) and Schreyer in general [1988], and even for some singular curves, have proved the conjecture, in a somewhat sharpened form, if Cliff C ≤ 2.

III. Special cases

Complete results are known -- and even published! -- in a few other cases. Most important, all possibilities for the resolutions of smooth canonical curves in every characteristic are known in case the genus is ≤ 8; this work was done by Schreyer in his Thesis [1986], and extended by him to certain singular curves in his [199?].

Loose has dealt with smooth plane curves of every genus, and also the smooth curves which can be embedded in $\mathbb{P}^3$ as the intersection of a surface and a quadric [1989]. It would be interesting to try to extend this work to some other complete intersections.

IV. Other approaches

A. Locally decomposable sections of Vector bundles:

A very different approach to Green's conjecture has been introduced by Paranjape and Ramanan [1988], who reduce it to some new and remarkably general questions about stable vector bundles on curves. The following remarks are due to them:

Consider a canonically embedded curve C ⊂ $\mathbb{P}^{g-1}$ (or the canonical map in the case of a hyperelliptic curve C) and let T be the vector bundle on $\mathbb{P}^{g-1}$ given by the obvious exact sequence

$$0 \to \mathcal{O}(-1) \to \mathcal{O}^g \to T \to 0$$

(so that T is the tangent bundle of $\mathbb{P}^{g-1}$ twisted by $\mathcal{O}(-1)$.) Let E be the restriction of T to C, and consider the natural map

$$p_j: \wedge^j H^0(E) \to H^0(\wedge^j E).$$

Paranjape and Ramanan show that this map is always an injection, and that Green's conjecture is equivalent to showing that it is a surjection for $j \le$ Cliff C.

Let $S \subset H^0(\wedge^j E)$ be the cone of "pointwise decomposable" sections: that is, sections σ such that at each point x of C, the element $\sigma_x \in E_x$ may be written as the exterior product of j elements of the vector space E_x. It is clear that the image of p_j, which is the span of all globally decomposable elements, must be contained in the span of S. Paranjape and Ramanan show that the image of p_j is actually equal to the span of S, the largest possible value, iff $j \le$ Cliff C.

There remains the problem of when the pointwise decomposable sections span the whole of $H^0(\wedge^j E)$. Paranjape and Ramanan have shown that they do span in the case of hyperelliptic curves, and with Hulek they show that they span if C is trigonal and $j = 2$ or C is plane quintic and j is arbitrary. But this question could be asked for <u>any</u> vector bundle E, generated by global sections, say:

Question: If E is a vector bundle on a curve C, generated by global sections, is it the case that the pointwise decomposable sections span $H^0(\wedge^j E)$ for every j? (The first open case would be $j=2$, rank E = 4.)

One would need only a much weaker result to settle Green's conjecture, as the bundle E of primary interest is stable (semi-stable if C is hyperelliptic) and has stronger and stronger stability properties as Cliff C increases.

B. The classification of cubics and the dual socle

Let R be the homogeneous coordinate ring of a canonical curve, and let x,y be general elements of R of degree 1. As we have explained above, $R/(x,y)$ is an Artinian Gorenstein ring with Hilbert function 1, γ, γ, 1, where $\gamma+2$ is the genus of C. We may of course write $R/(x,y) = k[z_1, \dots , z_\gamma]/J$. Such an Artinian Gorenstein factor ring corresponds to a cubic form F, up to scalars, in another set of variables $t_1, \dots t_\gamma$ as follows: If we regard the polynomial ring $k[t_1, \dots t_\gamma]$ as a module over $k[z_1, \dots , z_\gamma]$ by letting z_m act as the partial derivative $\partial/\partial t_m$ (this is in characteristic 0; in positive characteristic one must use divided power algebras...) then J is the annihilator of F and F is a generator of the submodule annihilated by J. (We are using the formulation of "inverse systems" of Macaulay; in modern terms, everything flows from the fact that $k[z_1, \dots , z_\gamma]$ is, with this module structure, the injective envelope of k as a $k[z_1, \dots , z_\gamma]$-module.)

It now makes sense to ask how the graded betti numbers in the free resolution of $R/(x,y)$ as a $k[z_1, \dots , z_\gamma]$-module, which are the same as the betti numbers in Green's conjecture, are reflected in the properties of the cubic form F. Cubic forms are interesting and much studied objects in their own right. In the case $\gamma=3$ (genus 5), which is the first not quite trivial case for Green's conjecture, the cubic forms in question correspond to cubic curves in the projective plane, an especially well-loved case. Michael Stillman and I (unpublished) have investigated this case, and found that the betti numbers of $R/(x,y)$ reflect quite a classical invariant of the cubic curve: ignoring the cases where on the one hand the cubic curve is a cone or on the other hand J contains a linear form, we found that the betti numbers are either of the form

```
1  -  -  -
-  3  -  -
-  -  3  -
-  -  -  1
```

(Cliff C = 2) or of the form

```
1  -  -  -
-  3  2  -
-  2  3  -
-  -  -  1
```

(Cliff C = 1.)

We proved that the latter case occurs precisely for cubics which are in the closure of the locus of smooth cubics of j-invariant 0 -- either the smooth cubic of j-invariant 0 or a cuspidal rational curve or the union of a conic and one of its tangent lines!

It would be interesting to extend the classification, even just to cubics in 4 variables.

References

E. Arbarello, M. Cornalba, P. A. Griffiths, and J. Harris. <u>Geometry of Algebraic Curves</u> vol I. (1985).

D. Bayer and D. Eisenbud. Ribbons. (in preparation)

D. Bayer and D. Eisenbud, appendix by S.-W. Park. Graph Curves. Appendix: Gonality and homoliferous connectivity of graph curves. Adv. in Math. (1991)

D. Bayer and M. Stillman, Some matrices related to Green's Conjecture. This Proceedings.

L. Ein. A remark on the syzygies of the generic canonical curves. Journal of Differential Geometry 26 (1987) 361--365.

D. Eisenbud. Linear Sections of Determinantal varieties. Am. J. Math. 110 (1988) 541-575.

D. Eisenbud. Transcanonical embeddings of hyperelliptic curves. J. Pure and Applied Alg. 19 (1980) 77-83.

D. Eisenbud and J. H. Koh. Nets of alternating matrices and the linear syzygy conjecture. Adv. in Math. to appear (1991?a)

D. Eisenbud and J.H. Koh. Some linear syzygy conjectures. Adv. in Math. to appear (1991?b)

F. Ferola. Resolutions of algebraic varieties. Unfinished thesis

L.-Y. Fong Studies on families of Curves. Brandeis Thesis (1991)

M. Green. Koszul cohomology and the geometry of projective varieties. (with an appendix by M. Green and R. Lazarsfeld.) J. Diff. Geom. 19 (1984) 125-171.

M. Green. Quadrics of rank four in the ideal of a canonical curve. Invent. Math. 75 (1984) 85--104.

M. Green and R. Lazarsfeld. Some results on the syzygies of finite sets and algebraic curves. Compositio Math. 67 (1988) 301-314.

M. Green and R. Lazarsfeld. A simple proof of Petri's theorem on canonical curves. Geometry today (Rome, 1984) Progr. Math., 60 Birkhauser Boston, Boston, Mass. (1985) 129-142.

M. Green and R. Lazarsfeld. On the projective normality of complete linear series on an algebraic curve. Invent. Math. 83 (1985) 73-90.

J. Koh and M. Stillman. Linear syzygies and line bundles on an algebraic curve. J. Alg. 125 (1989) 120-132.

A. Lascoux. Syzygies des variétés déterminantielles. Adv. in Math. 30 (1978) 202-237.

F. Loose. On the graded Betti numbers of plane algebraic curves. Manuscripta Mathematica 64 (1989) 503--514.

G. Martens and F.-O. Schreyer. Line bundles and syzygies of trigonal curves. Abhandlungen aus dem Mathematischen Seminar der Universitat Hamburg 56 (1986) 169--189.

K. Paranjape and S. Ramanan. On the canonical ring of a curve. Algebraic geometry and commutative algebra, Vol. II Kinokuniya, Tokyo (1988) 503--516.

F.-O. Schreyer. Syzygies of canonical curves and special linear series. Mathematische Annalen 275 (1986) 105--137.

F.-O. Schreyer. A standard basis approach to syzygies of canonical curves. J. Reine Angew., to appear (199?)

F.-O. Schreyer. Green's conjecture for general p-gonal curves of large genus. (to appear) (199?)

R. Smith and R. Varley. Deformations of singular points on theta divisors. Theta functions---Bowdoin1987, Part 1 (Brunswick, ME, 1987) Proc. Sympos. Pure Math., 49, Part 1 Amer. Math. Soc., Providence, RI, (1989) 571--579.

C. Voisin. Courbes tetragonales et cohomologie de Koszul. Journal für die Reine und Angewandte Mathematik 387 (1988) 111--121.

J. Weyman. Tangent developable of the rational normal curve and Mark Green's conjectures. (notes from 1988, available from Weyman, Dept of Math., Northeastern Univ., Boston, MA 02115)

J. Weyman. Equations of strata for binary forms. J.of Alg. 122 (1989) 244-249.

Some Matrices Related to Green's Conjecture

Dave Bayer [*]   Mike Stillman [†]

January 14, 1991

1 Introduction

In this note we present two matrices with the property that if either has full rank, for a specific g, then Green's conjecture would be true for generic curves of genus g.

For simplicity, we assume that $g \geq 5$. K denotes the base field which we assume has characteristic zero.

Throughout, we assume familiarity with Green's conjecture; see Eisenbud's writeup in these proceedings ([Eis 91]). We present only a sketch of our derivations, although proofs for the lemmas and propositions contained here can be readily supplied by the reader.

Both matrices arise from using Koszul cohomology on g-cuspidal rational curves in $\mathbf{P}^{g-1}$. The first matrix is derived by using an Artinian version of the equations of the tangent developable of the rational normal curve of degree g. After row reducing the appropriate matrix, one obtains a matrix which we denote below by $\Omega(g, \lfloor g/2 \rfloor)$. The advantage of this matrix is that it is relatively small, and we were able to use Macaulay ([BaSt 90]) to verify Green's conjecture through $g = 18$ by row reducing these matrices. It should be possible to extend these machine computations to somewhat higher g.

The second matrix, which we denote below by $\Pi(g, \lfloor g/2 \rfloor)$, is derived in a manner which might apply to other situations. Starting with a g-cuspidal

[*] Partially supported by the Alfred P. Sloan Foundation, by ONR contract N00014-87-K0214, and by NSF grant DMS-90-06116.

[†] Partially supported by the U.S. Army Research Office through the Mathematical Sciences Institute of Cornell University, and by NSF grants CCR-89-01061 and DMS-88-02276.

rational curve in $\mathbf{P}^{g-1}$, we write down the Koszul cohomology sequence, without knowing the equations of the variety. Instead, we pull-back the computation to $\mathbf{P}^1$. We obtain a free $K[s,t]$-module, such that the dimension of the space of sections which push forward to the cuspidal curve determines the dimension of a Koszul cohomology group. The condition for a section to push forward is that the residues at the cusps all be zero. We then choose appropriate bases and row reduce the resulting matrix to finally obtain a relatively simple matrix.

To date, we have been unable to compute the ranks of either of these matrices theoretically, except in special cases. We present them here so that hopefully someone else might be able to make further progress.

We denote by $\Lambda(p, m..n)$ the set

$$\{A = (a_1, \ldots, a_p) \mid m \leq a_1 < \ldots < a_p \leq n\}.$$

If $A \in \Lambda(p, 1..n)$ is an index set, and $a < b$ are elements not in A, define $\text{sign}(A, a, b)$ to be the sign of the permutation which sorts $A \cup \{a, b\}$.

2   The tangent developable of the rational normal curve

Fix an integer $g \geq 5$. Let $X \subset \mathbf{P}^g$ be the tangent developable of the rational normal curve of degree g in $\mathbf{P}^g$. For more information about these surfaces, see [Eis 91] or [Sch 83].

Let

$$\phi : \mathbf{P}^1 \longrightarrow \mathbf{P}^g$$

$$(s, t) \mapsto (y_0, \ldots, y_g) = (s^g, s^{g-1}t, \ldots, t^g)$$

be a parametrization of the rational normal curve. X is the image of the map

$$\phi : \mathbf{P}^1 \times \mathbf{P}^1 \longrightarrow \mathbf{P}^g$$

defined by

$$(s, t) \times (\alpha, \beta) \mapsto \frac{\partial \phi}{\partial s}\alpha + \frac{\partial \phi}{\partial t}\beta.$$

Let $J \subset S = K[y_0, \ldots, y_g]$ denote the ideal of X. The hyperplane sections of X are all g-cuspidal rational curves of genus g.

The following conjecture would imply Green's conjecture for general curves of genus g:

Conjecture 2.1 (Green) *Let J be the ideal of the tangent developable of the rational normal curve of degree g. Let $h = \lfloor g/2 \rfloor$. Then*

$$\mathrm{Tor}_h^S(S/J, k)_{h+1} = 0.$$

In this section, we present matrices $\Omega(g, h)$ with the property that this conjecture holds if $\Omega(g, \lfloor g/2 \rfloor)$ has maximal rank.

The following quantities appear in these matrices.

Definition 2.2 *For integers $0 \le i, j \le g - 3$, let*

$$c_{ij} := \begin{cases} \dfrac{(i+1)(j+1)}{i+j+1} & \text{if } i+j \le g-2 \\[2mm] \dfrac{(g-i-1)(g-j-1)}{2g-3-i-j} & \text{if } i+j \ge g-2. \end{cases}$$

Given integers $1 \le a < b \le g - 3$, and $1 \le k \le g - 3$, define

$$D(k, a, b) := c_{ka} c_{b,(k+a) \bmod (g-2)} - c_{kb} c_{a,(k+b) \bmod (g-2)}.$$

The rational numbers $D(k, a, b)$ are often zero.

Proposition 2.3 *If $1 \le a < b \le g - 3$, and $1 \le k \le g - 3$, then*

$$D(k, a, b) = 0 \iff \text{either } a + b + k \le g - 2 \quad \text{or}$$
$$a + b + k \ge 2g - 4 \quad \text{or}$$
$$a + b + 2k = 2g - 4.$$

Let $g \ge 5$, and $1 \le h \le g - 2$. We define a matrix Ω, or $\Omega(g, h)$, having $(g-2)\binom{g-3}{h-2}$ rows and $h\binom{g-2}{h+1}$ columns. The index set for the rows of $\Omega(g, h)$ is the set

$$\{(A, v) \mid A \in \Lambda(h-2, 1..g-3), 0 \le v \le g-3\}.$$

The index set for the columns is the set

$$\{(B, k) \mid B = (b_1 < \ldots < b_h) \in \Lambda(h, 1..g-3), k \ge b_1\}.$$

A combinatorial argument shows that the number of such indices is $h\binom{g-2}{h+1}$.

Definition 2.4 *Let* Ω, *or* $\Omega(g, h)$ *be the matrix having the above index sets for rows and columns, and whose entries are*

$$\Omega(g,h)_{(A,v),(B,k)} := \begin{cases} \text{sign}(A,a,b)D(k,a,b) & \text{if } B = A \cup \{a,b\} \text{ and} \\ & v \equiv k+a+b \pmod{g-2} \\ 0 & \text{otherwise} \end{cases}$$

By Proposition 2.3, if $a+b+k-(g-2)$ is not between 0 and $g-3$, then $D(k,a,b) = 0$. Each row labelled by (A,v), for $v = 0$, is completely zero, since if $k+a+b = g-2$, then $D(k,a,b) = 0$.

Theorem 2.5 *If $g \geq 3$, and $1 \leq h \leq g-2$, then*

$$\dim \text{Tor}_h(S/J, K)_{h+1} = h\binom{g-2}{h+1} - \text{rank}\, \Omega(g,h).$$

Therefore, Green's conjecture for general curves of genus g would follow if $\Omega(g, \lfloor g/2 \rfloor)$ were to have full rank. It should be noted that $\Omega(g,h)$ often does *not* have full rank if $h < \lfloor g/2 \rfloor$.

Example 2.6 *If $g = 8$, and $h = 3$, then $\dim \text{Tor}_3(S/J, K)_4 = 21$.* The dimension of the kernel of $\Omega(8,3) : Q^{45} \rightarrow Q^{25}$ is 21, *not* 20, as it would be if the map were of full rank.

The matrix $\Omega(g, h)$ is the direct sum of $g - 2$ matrices. In fact, define the degree of a multi-index A to be $\deg A = ((\sum_{a \in A} a) \bmod (g-2)) \in \mathbb{Z}/(g-2)\mathbb{Z}$, and $\deg(A, v) := v + \deg A \in \mathbb{Z}/(g-2)\mathbb{Z}$, and similarly for (B, k). With this notation, $\Omega(g, h)_{(A,v),(B,k)} = 0$ whenever $\deg(A, v) \neq \deg(B, k)$.

Definition 2.7 *For each $w \in \{0..g-3\}$, define $\Omega(g, h, w)$ to be the matrix with columns indexed by the set*

$$\{(B, k) : B = (b_1 < \ldots < b_h) \in \Lambda(h, 1..g-3), k \geq b_1, \deg(B, k) = w\},$$

having $\binom{g-3}{h-2}$ rows, and whose entries are

$$\Omega(g, h, w)_{A,(B,k)} := \Omega(g, h)_{(A, w - \deg A),(B,k)}.$$

$\Omega(g, h)$ *is the direct sum of these matrices, therefore*

Proposition 2.8

$$\text{rank } \Omega(g, h) = \sum_{w=0}^{g-3} \text{rank } \Omega(g, h, w).$$

We checked Green's conjecture for genera $g \leq 18$ by using Macaulay to verify that each of the matrices $\Omega(g, h, w)$ has full rank, for $g \leq 18$, $h = \lfloor g/2 \rfloor$, and $0 \leq w \leq g - 3$.

In the remainder of this section, we sketch the proof of Theorem 2.5

Proposition 2.9 (Schreyer [Sch 83]) *If $g \geq 5$, the ideal J is generated by the quadrics*

$$(i + j - 1)y_i y_j - ij y_{i+j-1} y_1 - (i + j - ij - 1)y_{i+j} y_0,$$

for $2 \leq i \leq j \leq g - 2$, and $i + j \leq g - 1$, and the quadrics

$$(2g - i - j - 1)y_i y_j - (g - i)(g - j)y_{i+j-g+1} y_{g-1} + (g - i - 1)(g - j - 1)y_{i+j-g} y_g,$$

for $2 \leq i \leq j \leq g - 2$, and $i + j > g - 1$.

Furthermore, $y_0, y_g, y_1 - y_{g-1}$ is a regular sequence of S/J.

Let $I \subset R = k[x_0, \ldots, x_{g-3}]$ be the ideal generated by the $\binom{g-2}{2}$ quadrics

$$x_i x_j - c_{ij} x_0 x_{(i+j) \bmod (g-2)} \qquad \text{for each } 1 \leq i \leq j \leq g - 3,$$

I is obtained from the ideal $J \subset S$ which defines the tangent developable, by setting $y_0 = y_g = y_1 - y_{g-1} = 0$, and renaming the variables $x_0 := y_1 = y_{g-1}$, and $x_i := y_{i+1}$, for $1 \leq i \leq g - 3$.

Since $\{y_0, y_g, y_1 - y_{g-1}\}$ is a regular sequence of S/J, the graded betti numbers of S/J and R/I are exactly the same.

$\text{Tor}_h(R/I, K)_{h+1}$ is the homology of the complex

$$\bigwedge^{h+1} R_1 \xrightarrow{\gamma} \bigwedge^h R_1 \otimes R_1 \xrightarrow{\partial} \bigwedge^{h-1} R_1 \otimes (R/I)_2,$$

where R_d is the vector space of elements of degree d in R.

Let $V = R_1$. $(R/I)_2$ is isomorphic to V, and the multiplication map $R_1 \otimes R_1 \to (R/I)_2$ becomes $m : V \otimes V \to V$, where $m(x_i \otimes x_j) = c_{ij} x_{(i+j) \bmod (g-2)}$. The above complex transforms into

$$\bigwedge^{h+1} V \xrightarrow{\gamma} \bigwedge^h V \otimes V \xrightarrow{\partial} \bigwedge^{h-1} V \otimes V,$$

where γ is the usual Koszul map, and ∂ is the composite of the Koszul map and the multiplication m.

We introduce coordinates in order to write down the matrix $\Omega(g, h)$. Let $\{e_A \mid A = (a_1, \ldots, a_p) \in \Lambda(p, 0..g-3)\}$ be a basis of $\bigwedge^p V$, for each p, where $e_A = x_{a_1} \wedge \ldots \wedge x_{a_p}$.

The image of

$$\{e_B \otimes x_j \mid B \in \Lambda(h, 0..g-3), b_1 \leq j \leq g-3\}$$

in $\bigwedge^h V \otimes V / \operatorname{Im} \gamma$ is a basis of this vector space, since $\gamma(e_{B \cup \{i\}}) = e_B \otimes x_i +$ lower terms, for each basis vector $e_{B \cup \{i\}} \in \bigwedge^{h+1} V$, where $i < b_1$.

We now row reduce the matrix corresponding to

$$\partial : \bigwedge^h V \otimes V / \operatorname{Im} \gamma \longrightarrow \bigwedge^{h-1} V \otimes V.$$

If $J \in \Lambda(h-1, 1..g-3)$, and $0 \leq k \leq g-3$, then

$$\partial(e_{J \cup \{0\}} \otimes x_k) = e_J \otimes x_k - \sum_{\alpha=1}^{h-1} (-1)^{\alpha+1} c_{j_\alpha, k} e_{J \cup \{0\} \setminus j_\alpha} \otimes x_{(j_\alpha+k) \bmod (g-2)}.$$

We reduce the other basis elements using these. If $B \in \Lambda(h, 1..g-3)$, and $b_1 \leq k \leq g-3$, then

$$\partial(e_B \otimes x_k) = \sum_{\beta=1}^{h} (-1)^{\beta+1} c_{b_\beta, k} e_{B \setminus b_\beta} \otimes x_{(b_\beta+k) \bmod (g-2)}.$$

Therefore

$$\partial(e_B \otimes x_k) - \sum_{\beta=1}^{h} (-1)^{\beta+1} c_{b_\beta, k} \partial(e_{B \cup \{0\} \setminus b_\beta} \otimes x_{(b_\beta+k) \bmod (g-2)})$$

$$= \sum_{1 \leq \alpha < \beta \leq h} (-1)^{\alpha+\beta+1} D(k, b_\alpha, b_\beta) e_{B \cup \{0\} \setminus \{b_\alpha, b_\beta\}} \otimes x_{(b_\alpha+b_\beta+k) \bmod (g-2)}.$$

$\Omega(g, h)$ is obtained as the matrix corresponding to these elements. $\square$

3   Cuspidal curves

A rational projective curve in $\mathbf{P}^{g-1}$ having g cusps and degree $2g-2$ is projectively equivalent to the image of a map

$$\mathbf{P}^1 \longrightarrow \mathbf{P}^{g-1}$$

defined by

$$(s, t) \mapsto (\frac{1}{(s - z_0 t)^2}, \ldots, \frac{1}{(s - z_{g-1} t)^2}),$$

where $z_0, \ldots, z_{g-1} \in K$. Without loss of generality, we set $z_0 = 0$.

The homogeneous coordinate ring $R(C)$ of the image C of the above map is isomorphic to $R = K[\theta/\theta_0, \ldots, \theta/\theta_{g-1}]$, where $\theta := \prod_{i=0}^{g-1}(s - z_i t)^2$, and for $i = 0, \ldots, g-1$, $\theta_i := (s - z_i t)^2$. If $A \subset \{0..g-1\}$, define polynomials $\theta_A := \prod_{a \in A}(s - z_a t)^2$.

Green's conjecture for generic curves follows from the following conjecture, which throughout this note we also refer to as Green's conjecture.

Conjecture 3.1 (Green) *Let* $S = K[x_0, \ldots, x_{g-1}]$, *and write* $R(C) = S/I$, *where* C *is the cuspidal curve defined above. If* $h = \lfloor g/2 \rfloor$, *then*

$$\dim \mathrm{Tor}_h^S(S/I, K)_{h+1} = 0.$$

Remark 3.2 If Conjecture 3.1 is true for $g = 2h + 1$, then the conjecture also holds for $g = 2h$. In fact, if $z_0, \ldots, z_{g-1}, \ldots, z_{g'-1}$ are elements of K, and if $I_g \subset K[x_0, \ldots, x_{g-1}]$ is the ideal of the image of the map above using values $z_0, \ldots, z_{g-1}$, then for every $g < g'$, $I_g \subset I_{g'}$. Therefore the 2-linear strand of the resolution of I_g injects into the 2-linear strand of the resolution of $I_{g'}$, for every $g < g'$.

Consequently, we can restrict to odd values of g.

Definition 3.3 *Fix* $1 \leq h \leq g-1$. *Define a matrix* Π *(or* $\Pi(g, h)$*) having* $\binom{g-1}{h-1}(g - h)$ *rows and* $\binom{g-1}{h+1}(h + 1)$ *columns as follows.*

The rows of Π *are indexed by* $\{(A, m) \mid A \in \Lambda(h-1, 1..g-1), m \notin A\}$, *and the columns of* Π *are indexed by* $\{(B, m') \mid B \in \Lambda(h + 1, 1..g - 1), m' \in B\}$.

If $B = A \cup \{a, b\}$, *where* $a < b$, *define*

$$\Pi_{\{A,m\},\{B,m'\}} = \begin{cases} \dfrac{\mathrm{sign}(A, a, b)}{(z_a - z_b)^3} & \text{if } m = m', \text{ and } m \in \{a, b\} \\[3mm] -\dfrac{\mathrm{sign}(A, a, b)}{(z_a - z_b)^3} & \text{if } m \neq m', \text{ and } m, m' \in \{a, b\} \\[3mm] 0 & \text{in all other cases} \end{cases}$$

If $A \not\subset B$, *then define*

$$\Pi_{\{A,m\},\{B,m'\}} = 0.$$

II is a block matrix with $\binom{g-1}{h-1}$ by $\binom{g-1}{h+1}$ blocks each of size $(g-h)\times(h+1)$. Each block is a scalar multiple of a matrix with entries in the set $\{0,1,-1\}$. The sum of every row in a block is zero, and consequently the maximum rank of II is $\binom{g-1}{h-1}(g-h-1)$.

Example 3.4 For $g = 5$, and $h = 2$, the matrix II is presented below. We have abbreviated $1/(z_p - z_q)^3$ by $[pq]$. We have also bordered the matrix with the row and column indices.

		123			**124**			**134**			**234**		
		1	2	3	1	2	4	1	3	4	2	3	4
	2		[23]	-[23]		[24]	-[24]						
1	3		-[23]	[23]					[34]	-[34]			
	4					-[24]	[24]		-[34]	[34]			
	1	-[13]		[13]	-[14]		[14]						
2	3	[13]		-[13]								[34]	-[34]
	4				[14]		-[14]					-[34]	[34]
	1	[12]	-[12]					-[14]		[14]			
3	2	-[12]	[12]								-[24]		[24]
	4							[14]		-[14]	[24]		-[24]
	1				[12]	-[12]		[13]	-[13]				
4	2				-[12]	[12]					[23]	-[23]	
	3							-[13]	[13]		-[23]	[23]	

Theorem 3.5 *Fix* $1 \le h \le g - 2$. *Then*

$$\dim \operatorname{Tor}_h^S(S/I, K)_{h+1} = \binom{g-1}{h+1}h - \operatorname{rank} \text{II}.$$

In the case $g = 2h + 1$ is odd, this is zero if and only if $\operatorname{rank} \text{II} = \binom{g-1}{h-1}(g - h - 1)$, which by the remark after Definition 3.3 is the maximum rank possible.

In the remainder of this section, we sketch the derivation of Theorem 3.5. Let $V = S_1 = R_{2g-2}$. If

$$\partial : \bigwedge^h V \otimes R_{2g-2} \longrightarrow \bigwedge^{h-1} V \otimes R_{4g-4}$$

is the Koszul map, then $\dim \operatorname{Tor}_h(S/I, K)_{h+1} = \dim \ker \partial - \binom{g}{h+1}$.

The technique we use is to pullback this map to $\mathbf{P}^1$, in the following sense. There is a diagram

$$
\begin{array}{ccccc}
\ker \partial & \subset & \bigwedge^h V \otimes R_{2g-2} & \xrightarrow{\partial} & \bigwedge^{h-1} V \otimes R_{4g-4} \\
\downarrow & & \downarrow & & \downarrow \\
\ker \Delta & \subset & \bigwedge^h V \otimes K[s,t]_{2g-2} & \xrightarrow{\Delta} & \bigwedge^{h-1} V \otimes K[s,t]_{4g-4},
\end{array}
$$

where each vertical arrow is an inclusion. Given such a diagram, one can compute the dimension of $\ker \partial$ using the following trivial lemma.

Lemma 3.6 *Given vector spaces $W_1, W_2 \subset W$, let $W_2^\perp \subset W^*$ be the annihilator of W_2 in W. If $\Phi : W_1 \to (W_2^\perp)^*$ is the natural map, then $W_1 \cap W_2 = \ker \Phi$.*

The derivation below proceeds in the following way. We first compute $\ker \Delta$ (Lemma 3.7), and choose a convenient basis (Lemma 3.8). In Lemma 3.9, we compute elements of the dual, which cut out the subspace $\bigwedge^h V \otimes R_{2g-2}$. These elements are essentially the residues of functions at the g cusps. Given this basis and generating set, we obtain a matrix whose kernel is the same as $\ker \partial$. In order to determine the rank of this matrix, there are obvious rows and columns which can be removed. The result is the matrix Π defined above. Theorem 3.5 follows immediately given this derivation.

Lemma 3.7 *Let V and W be free $K[s,t]$-modules of ranks g and $g-1$, respectively. There exists an exact sequence of $K[s,t]$-modules*

$$0 \to \bigwedge^h W \xrightarrow{\gamma} \bigwedge^h V \xrightarrow{\Delta} \bigwedge^{h-1} V,$$

where Δ is the Koszul map defined above, and γ is defined on a basis vector e_C, $C = (0 = c_1, c_2, \ldots, c_{h+1}) \in \Lambda(h+1, 0..g-1)$ containing zero by

$$\gamma(e_C) = \sum_{i=1}^{h+1} (-1)^{i+1} \theta_{C \setminus c_i} e_{C \setminus c_i}.$$

Lemma 3.8 *The elements*

$$(i) \quad \frac{\theta}{\theta_C} e_C$$

$$(ii) \quad \frac{\theta}{\theta_C} \frac{(s - z_{c_2} t)^2}{(s - z_l t)^2} e_C, \qquad l \notin C,$$

$$(iii) \quad \frac{\theta}{\theta_C} \frac{s^2}{(s - z_l t)^2} e_C, \qquad l \notin C,$$

where $C \in \Lambda(h+1, 1..g-1)$ contains 0, generate the vector space $\bigwedge^h W \otimes K[s,t]_{2g-2-2h}$.

This lemma is easily checked by comparing residues at $s = z_l t$, after dividing by θ/θ_C. The element c_2 in the second set of elements of the basis can be chosen to be any (non-zero) element of C, not necessarily the first non-zero element.

The elements of the dual of $\bigwedge^h V \otimes R_{2g-2}$ are determined by the following lemma.

Lemma 3.9 *A polynomial $f \in K[s,t]_{2g-2}$ lies in the subspace R_{2g-2} if and only if the residues $\operatorname{Res}_{s=z_l} f(s,1)/\theta(s,1) = 0$, for every $0 \le l \le g-1$.*

Note that the sum of the residues is zero, and therefore these g conditions only determine $g-1$ independent conditions.

The (dual of the) annihilator of $\bigwedge^h V \otimes R_{2g-2}$ in $\bigwedge^h V \otimes k[s,t]_{2g-2}$ is generated by elements $\operatorname{Res}_{A,m}$ for $A \in \Lambda(h, 0..g-1)$, and $m = 0, \ldots, g-1$, where

$$\operatorname{Res}_{A,m}(f(s,t)e_C) = \operatorname{Res}_{s=z_m} (f(s,1)/\theta(s,1))\langle e_A, e_C \rangle,$$

and

$$\langle e_A, e_C \rangle = \begin{cases} 0 & \text{if } A \ne C \\ 1 & \text{if } A = C \end{cases}.$$

Lemma 3.10 *The residue*

$$\operatorname*{Res}_{s=c} \frac{(s-a)^2}{(s-b)^2(s-c)^2} = \frac{2(a-c)(a-b)}{(b-c)^3}.$$

Consider the linear map

$$\Phi : \bigwedge^h W \otimes K[s,t]_{2g-2-2h} \longrightarrow (\bigwedge^h V \otimes R_{2g-2})^{\perp \bullet}$$

obtained by applying Lemma 3.6. Note that the images of the $\binom{g-1}{h}$ basis elements of Lemma 3.8 of type (i) are all zero. In fact,

$$\operatorname*{Res}_{A,m} \gamma(\theta/\theta_C \, e_C) = \sum_{i=1}^{h+1}(-1)^{i+1} \operatorname*{Res}_{s=z_m} \frac{1}{(s-z_{c_i})^2}\langle e_A, e_{C \backslash c_i}\rangle = 0,$$

since all such residues are zero.

Note that, for fixed A not containing zero, if $C = A \cup \{0\}$, then

$$\operatorname*{Res}_{A,l} \gamma(\frac{\theta}{\theta_C} \frac{(s-z_{c_2}t)^2}{(s-z_l t)^2} e_C) \ne 0,$$

whereas $\text{Res}_{A,l}$ applied to any other basis element of Lemma 3.8 of type (ii) or type (iii) results in zero. Therefore, to compute $\ker \Phi$, we can ignore every basis element of Lemma 3.8 of type (ii), and only consider those rows which correspond to index sets $A \subset \Lambda(h, 0..g-1)$ which contain zero.

If $A \subset \{1..g-1\}$ is an index set, let $z_A := \prod_{a \in A} z_a$. The matrix Π is obtained in a straightforward manner by using

$$Y_{B,m'} = (-1)^j \frac{\theta}{z_B \theta_B} e_{B \setminus m' \cup \{0\}},$$

for $B \in \Lambda(h+1, 1..g-1)$, $m' = b_j \in B$, as a basis for the column space, and

$$X_{A,m} = \frac{z_A}{2} \text{Res}_{A \cup \{0\}, m}$$

for $A \in \Lambda(h-1, 1..g-1)$, $m \notin A, 1 \leq m \leq g-1$, as a generating set for the row space.

Finally, Theorem 3.5 follows immediately:

$$\dim \text{Tor}_h(S/I, K)_{h+1} = \dim \ker \theta - \binom{g}{h+1}$$

$$= \dim \ker \Phi - \binom{g}{h+1}$$

$$= \binom{g-1}{h} + \binom{g-1}{h+1}(h+1) - \text{rank}\,\Pi - \binom{g}{h+1}$$

$$= \binom{g-1}{h+1}h - \text{rank}\,\Pi,$$

as desired.   □

References

[BaSt 90]   D. Bayer and M. Stillman, *Macaulay: A system for computation in algebraic geometry and commutative algebra*, available for Unix and Macintosh computers. Contact the authors, or ftp zariski.harvard.edu, Name: ftp, Password: any, cd Macaulay, binary, get M3.tar, quit, tar xf M3.tar.

[Eis 91]   D. Eisenbud, *Green's conjecture: An orientation for algebraists* These proceedings (1991).

[Gr 84]   M. Green, *Koszul cohomology and the geometry of projective varieties*, (with an appendix by M. Green and R. Lazarsfeld). J. Diff. Geom. 19, 125–171 (1984)

[Sch 86]   F.-O. Schreyer, *Syzygies of canonical curves and special linear series*, Math. Annalen 275, 105–137 (1986).

[Sch 83]   F.-O. Schreyer, *Syzygies of curves with special pencils* Ph.D. Thesis, Brandeis (1983).

Other Topics

Problems on Local Cohomology

CRAIG HUNEKE*

1. Introduction.

Let R be a noetherian ring, let I be an ideal of R, and let M be an R-module. The local cohomology of M with respect to I is by definition

$$\varinjlim_{\ell} \operatorname{Ext}_R^n(R/I^\ell, M) = H_I^n(M).$$

As is well-known, these local cohomology modules can also be computed in the case that $M = R$ by using a Cech resolution of $\operatorname{Spec}(R) - V(I)$, and consequently relate to the topology of these open sets. In this paper we will discuss results and questions centered about four basic problems concerning local cohomology.

Problem 1. When are $H_I^n(M) = 0$?

Problem 2. When are $H_I^n(M)$ finitely generated?

Problem 3. When are $H_I^n(M)$ Artinian?

Problem 4. If M is finitely generated, is the number of associated primes of $H_I^n(M)$ always finite?

We will find that all of these problems are connected with another question: what annihilates the local cohomology $H_I^n(M)$?

2. Problem 1.

There has been a great deal of work on this problem. The first and most basic theorem was given by Grothendieck himself

---

*Partially supported by the NSF

who showed that $H_I^n(M) = 0$ if $n > \dim(R)$. He also showed that the dth local cohomology of a local ring of dimension d at its maximal ideal is never zero. In addition, $H_I^n(M) = 0$ if n is bigger than the least number of equations needed to define I up to radical, ara(I). We can formalize the least number at which such vanishing occurs by defining the *cohomological dimension of R with respect to I*, $cd(I, R)$, = greatest integer n such that there exists a module M with $H_I^n(M) \neq 0$. It is not difficult to show that this integer is the same as the largest integer n such that $H_I^n(R) \neq 0$. For this reason, we turn our attention to the case where $M = R$. As flat base change commutes with local cohomology we may further assume that R is complete and local. The cohomological dimension is the same as the maximum for R modulo its minimal primes, so there is no loss in generality in assuming that R is a complete local domain of dimension d, say.

If R is a complete local domain of dimension d and if $\dim(R/I) > 0$, Hartshorne [H1] proved that $cd(I, R) < d$ (see also [BH], [CS], [PS]). This result is very useful; for instance Brodmann and Rung [BR] (cf. [F3] also) use it to show that if (R, m) is a complete local domain of dimension d, and $a_1, \ldots, a_k$ are in m with $k \leq d - 2$, then the punctured spectrum of R/J is connected, where J is generated by the a_i. Recently Hochster and Huneke have been able to generalize this result by weakening the assumption that R is a domain to just assuming that the canonical module of R is indecomposable. This not only includes domains, but also includes the case where R is S_2.

The natural question to ask next is when is $cd(I, R) < d - 1$? A necessary condition is that $\dim(R/I) \geq 2$; for if p is a minimal prime over I having height h, then

$H_I^n(R) \neq 0$, since it is not zero after localizing at p. This necessary condition is not enough. Basically what can go wrong is that a disconnection of the punctured spectrum will cause an increase in the cohomological dimension. This is all that can go wrong in the case R is regular. The following theorem was proved by Hartshorne in the geometric case, by Ogus in characteristic 0 [O], and by Peskine-Szpiro [PS] and Hartshorne-Speiser [HS] in characteristic $p > 0$. Recently a characteristic-free proof was given by Huneke-Lyubeznik [HuL].

THEOREM 2.1. *Let R be a complete regular local ring of dimension d with a separably closed residue field. Let I be an ideal of R. The following are equivalent:*

i) $cd(I, R) < d - 1$.

ii) $\dim(R/I) \geq 2$, *and the punctured spectrum of R/I is connected.*

There can be great differences between the theory in characteristic p and 0 as was shown by Peskine and Szpiro. The flatness of the Frobenius map over regular rings of characteristic p allows one to show that if R/I is Cohen-Macaulay, where R is a regular local ring of characteristic p, then $cd(I, R) = ht(I)$. This is totally false in characteristic 0. The simplest example is the 2 by 2 minors of a generic 2 by 3 matrix over the complex numbers. The cohomological dimension is 3, although the height is 2 and the ideal defines a Cohen-Macaulay ring.

Unfortunately, there is no simple extension of the above theorem to lower cohomological dimension, or to non-regular rings. However during the seventies, work continued, especially centered about cohomological bounds for the ideals in rings which were complete intersections, for abelian varieties, or for ideals I which are complete intersections on the punctured spectrum cf. [S1-3] for example. In 1980, Faltings [F1] gave the most general vanishing result:

THEOREM 2.2. *Let R, m be a complete local ring containing a field and write $R = S/I$, where S is a regular local ring. Let J be an ideal of R, and lift back J to J' in S. Set $b = \text{bight}(J')$ (which is the biggest height of any minimal prime of J'). Put $d = \dim(R)$. Then $cd(J, R) \leq d - [(d-1)/b]$.*

Lyubeznik [Ly] has shown that this result is the best possible in general. Any further progress is tied to assuming more about the structure of R/I. For instance the case where R/I has an isolated singularity was treated by Speiser, Barth, and Faltings. In [HuL] more was shown in the case that J is either a prime ideal or R/J is normal.

THEOREM 2.3. *[HuL] Let the notation be as in the theorem above.*

i) If J is formally geometrically irreducible, then

$$cd(J, R) \leq d - 1 - [(d-2)/b].$$

ii) If R/J is normal, then

$$cd(J, R) \leq d - [(d+1)/(b+1)] - [d/(b+1)].$$

In *[HuL] it is shown that i) is the best possible, and at least for some values of b and d that ii) is the best possible.*

We believe that i) should hold even in the case J is not prime, provided that J is not the intersection of too many primes. Precisely:

CONJECTURE 2.4. *Let R be a complete regular local ring with separably closed residue field, and let I be an ideal such that the nilradical of I is the intersection of t primes of maximal height h. If $th + 1 \leq d$ then $cd(I, R) \leq d - 1 - [(d-2)/h]$. The condition that $th + 1 \leq d$ guarantees that the minimal primes of I cannot add to an m-primary ideal.*

Although there is no formulated conjecture, three general problems become evident.

Problem 2.5. Assume that R/I satisfies Serre's property S_i and R_j. What is the maximum possible cohomological dimension for such an ideal?

Evidently the answer in general will be different in different characteristics.

Problem 2.6. Suppose that I is generated by t elements. When is $H_I^t(R) = 0$?

This question arises in the proof of the theorems above and in other contexts. It seems to be very difficult in general to give any criterion for this vanishing (other than a rephrasing). The most important case is to understand when the *diagonal* has this vanishing property. Let R be a power series ring in n-variables over a field, and let I be an ideal of R. Let S be the complete tensor product of R with R/I, and let J be generated by the diagonal in S. Then, $cd(J, S) \leq n$, the number of generators of J. All of the results in Theorem 2.3 can be improved if a better bound could be given for $cd(J, S)$.

Problem 2.7. This problem arises with the computation of local cohomology. Let R be a polynomial ring over a field and let I be a homogeneous ideal of R. How can we compute (i.e., using Macaulay, or some other computer algebra program) pieces of the local cohomology modules $H_I^n(R)$? (Or in general the local cohomology of any finitely generated graded R-module.) Unfortunately the graded pieces of this local cohomology need not be finite-dimensional. However suppose we fix a degree k and a fixed power n of I and look at the elements in the local cohomology in degree k and annihilated by I^n. Is this finite dimensional? If so, how can one calculate it? It is a direct limit of objects which can be calculated, so the only question is to give bounds on how far out in the direct limit one needs to go to make sure one has covered the piece of the local cohomology in question.

3. When is the local cohomology finitely generated?

The simplest case of this question is when the ideal in question is the maximal ideal of a local Cohen-Macaulay ring; in this case the local cohomology of a finitely generated module M is automatically Artinian and hence is finitely generated iff it has finite length. In this case by using local duality, one can convert this into a statement about the support of the Ext modules of M into the canonical module of $\hat{R}$, and this reduction answers the question as well as it can be answered; one can examine the local depths of the module.

Faltings [F2] gave a fairly complete answer to this question in general; at least he identified the smallest local cohomology which is not finitely generated. To do this he first proved the following nice proposition which related the finite generation property to the annihilators of local cohomology.

PROPOSITION 3.1. *[F2] Let R be a noetherian ring and let I be an ideal, M a finitely generated R-module. The following are equivalent:*

i) $H_I^j(M)$ are finitely generated for all $j < n$.

ii) $H_I^j(M)_p$ are finitely generated for all p in $\mathrm{Spec}(R)$ and for all $j < n$.

iii) There exists a power k of I such that $I^k(H_I^j(M)) = 0$ for all $j < n$.

Using this proposition Faltings gave a criterion for finite generation which depends on the numbers,

$$s(I, M) = \min\{\mathrm{depth}(M_p) + ht((I + p)/p) : p \not\supseteq I\}.$$

THEOREM 3.2. [F2] If $j < s(I, M)$, then $H_I^j(M)$ is finitely generated. If $j = s(I, M)$, then $H_I^j(M)$ is not finitely generated.

Brodmann [B] has generalized this to give a criterion for when $H_I^j(M)$ is annihilated by a power of some ideal J contained in I. Recently Raghavan [R] proved that the power of J in Brodmann's theorem needed to annihilate $H_I^j(M)$ is independent of both I and J, depending only on the module M. This result has interesting consequences. A wild problem is the following:

Problem 3.3. Let M be a finitely generated R-module. Consider the set of integers,

$$W = \{\mathrm{depth}(M_p) + ht(I + p)/p, \text{ where } p \not\supseteq I\}.$$

Then is $n \geq 0$ not in W iff $H_I^n(M)$ is finitely generated?

Grothendieck [G2] conjectured the following weaker statement which would give that while the local cohomology need not be finitely generated, a large piece of it is:

CONJECTURE 3.4. *[G2]* $\mathrm{Hom}_R(R/I, H_I^j(R))$ *is finitely generated for all j and all I.*

This conjecture was shown to be false by Hartshorne in [H2]. His example is the following. Let $R = k[x, y, u, v]/(xy - uv)$, and let $I = (x, y)$. Then $H_I^2(R)$ has an infinite dimensional socle, and in particular, $\mathrm{Hom}_R(R/I, H_I^2(R))$ must be non-finitely generated. However, the base ring R in Hartshorne's example is not regular and the problem was still open for regular rings. However, a recent result in [HuK] shows these Hom's are almost never finitely generated.

THEOREM 3.5. *[HuK] Let R be a regular local ring, I an ideal of R, and let $b = \mathrm{bight}(I) = $ biggest height of any minimal prime of I.*

a) *If characteristic $R = p > 0$, and if $n > b$ is such that $\mathrm{Hom}_R(R/I, H_I^n(R))$ is finitely generated, then $H_I^n(R) = 0$.*
b) *If R contains the rationals, and if $n > b$ is the largest integer such that $H_I^n(R)$ is nonzero, then $\mathrm{Hom}_R(R/I, H_I^n(R))$ is never finitely generated.*

In both [H2] and [HuK] some cases of finite generation for the Hom were given, but in general the picture looks somewhat bleak—these local cohomology modules are quite big if they are non-zero, at least past the big height of the ideal. A specific example in characteristic 0 of this phenomena is given by letting I be the 2 by 2 minors of a generic 2 by 3 matrix over the complex numbers. If R is the ambient polynomial ring, then $H_I^3(R) \neq 0$. This was shown by Hochster. It is also the largest nonvanishing local cohomology in this case as I has only 3 generators, and of course the height of $I = \mathrm{bight}(I)$ is 2. Thus $\mathrm{Hom}_R(R/I, H_I^3(R))$ must be infinitely generated.

Given these facts, it is somewhat amazing that there is still hope that the local cohomology modules of an ideal in a regular ring behave a great deal like a finitely generated module. This is best explained in the context of trying to decide when the local cohomology is Artinian.

4. When is the local cohomology Artinian?

We first recall some general facts. Let (R, m, k) be a noetherian local ring and let M be an R-module. If M is Artinian then $\mathrm{Supp}(M) = \{m\}$, and the socle of M, denoted $\mathrm{soc}(M)$, $= \mathrm{Hom}_R(k, M)$ is finitely generated. Conversely, these two conditions imply that M is Artinian. To say that $\mathrm{Supp}(M) = \{m\}$ is equivalent to saying that M is an essential extension of its socle. Hence if $\mathrm{soc}(M)$ is finitely generated, M will be a submodule of a finite direct sum of copies of the injective hull of the residue field, and thus be Artinian. This breaks the problem into two pieces; when is the local cohomology supported at the maximal ideal, and when is the socle finitely generated? The first problem can be restated to ask when is the localization of a local cohomology module zero at every prime ideal which is not the maximal ideal. In this form, this is a question of the vanishing of the local cohomology, and can be treated as we discussed in the first section. Therefore in this section we will chiefly consider the second problem. Two results concerning when the local cohomology are Artinian were proved in the seventies, the characteristic 0 version due to Ogus, and the characteristic p version due to Hartshorne and Speiser.

THEOREM 4.1. [O] Let (R, m, k) be a regular local ring containing the rationals. Let I be an ideal of R, and suppose that for all $n > j$, $H_I^n(R)$ has support only at m. Then for all $n > j$, $H_I^n(R)$ is Artinian.

THEOREM 4.2. *[HS] Let (R, m, k) be a regular local ring of characteristic $p > 0$. Let I be an ideal of R, and suppose that for some n, $H_I^n(R)$ has support only at m. Then $H_I^n(R)$ is Artinian.*

These theorems are similar to the result of [HuK] above in that for the prime characteristic case, one can say something about individual local cohomology, while for the equicharacteristic 0 case, one must assume something for all local cohomology modules past some value. However I believe that the hypothesis that the support of the module is m is a red herring, and what should generally be true is that the socle should always be finitely generated. I can formalize this in a conjecture:

CONJECTURE 4.3. *Let (R, m, k) be a regular local ring and let I be an ideal of R. For all n, $\text{soc}(H_I^n(R))$ is finitely generated.*

There is some evidence that this is true. In [HuS], it will be shown that conjecture 4.3 holds for regular local rings in characteristic $p > 0$. In equicharacteristic 0, or in mixed characteristic, this problem is open. It is somewhat difficult to approach this problem as first of all it is definitely not true for modules in place of the ring R as Hartshorne's example has shown. Moreover, it is not a very stable assumption. For instance in a short exact sequence, if two of the socles are finitely generated, it does not necessarily force the third socle to be finitely generated due to the presence of an Ext^1. This can be overcome by hoping that $\text{Ext}_R^i(k, H_I^n(R))$ be finitely generated for all i. A more succinct way of saying this, which would show the local cohomology exhibits characteristics of finitely generated modules is the following conjecture, more general than (4.3):

CONJECTURE 4.4. *Let (R, m, k) be a regular local ring. Let I be an ideal of R. Then the Bass numbers $\text{Ext}^i_{R_p}(k(p), H^n_{I_p}(R_p))$ are finite for all i and n. In particular the injective resolution of the local cohomology has only finitely many copies of the injective hull of R/p for any p.*

Again there is evidence that this conjecture is true. It will be shown in [HuS] that the conjecture holds in positive characteristic. The first interesting case in equicharacteristic 0 is the case of a height 2 ideal in dimension 5. For smaller dimension or bigger heights the theorems on vanishing together with the theorem of Ogus above show that the conjecture is true. At present, it seems like these conjectures represent the best one can expect in general from the local cohomology modules. One more type of good behavior which is possible is the subject of the last section.

5. Are there only finitely many associated primes?

In general form, it seems possible to make the following strong conjecture.

CONJECTURE 5.1. *Let (R, m) be a noetherian local ring, let I be an ideal of R, and M be a finitely generated R-module. Then for all $j \geq 0$, the number of associated primes of $H^j_I(M)$ is finite.*

In this form, the conjecture is completely open. A more plausible conjecture might be,

CONJECTURE 5.2. *Let (R, m) be a regular local ring, and let I be an ideal of R. Then for all $j \geq 0$ $H^j_I(R)$ has only finitely many associated primes.*

The second conjecture has recently been proved in characteristic p in [HuS]. Unfortunately, there does not seem to be any hope of using reduction to characteristic p when it comes to the vanishing of elements in local cohomology. Nonetheless, this result in characteristic p gives hope in all other cases. Another suggestive result was given in the proof of Faltings of Proposition 3.1. One of the key steps in the proof in showing the modules were finitely generated was to show that the number of associated primes was finite by an induction. Of course, after the fact this is true in Faltings' case since he was showing the modules were actually finitely generated.

On the negative side the result of Huneke and Koh (3.5) shows that even $\operatorname{Hom}_R(R/I, H_I^n(R))$ will not be finitely generated for $n > \operatorname{bight}(I)$ in a regular local ring. While this does not mean there are infinitely many associated primes, it certainly gives one pause.

An interesting problem in this connection, first raised as far as I know by Markus Brodmann, is the following question.

Question 5.3. Let R be a noetherian ring and let I and J be ideals of R. Fix an integer $j \geq 0$. Suppose for each maximal ideal m of R, J_m is contained in the nilradical of the annihilator of $H_{I_m}^j(R_m)$. Then is J contained in the nilradical of $H_I^j(R)$?

This question would have an immediate positive solution if the local cohomology module had only finitely many associated primes. There has been some progress recently on this question by Raghavan.

Another somewhat whimsical question I raised some years ago is the following.

Question 5.4. Let (R, m) be a noetherian local ring and let $x_1, \ldots x_n$ be any collection of elements of R. Consider the set of all associated primes of all ideals generated by monomials in the x_i. Is this a finite set?

Of course, the well-known result of M. Brodmann saying that the number of associated primes of powers of ideals is a finite set would be an immediate corollary of this question. The only classes of elements for which I know a positive solution are when the x_i form a regular sequence, or when $n = 1$, both being rather trivial cases. Very likely it is true if the x_i form a d-sequence, and it might be a good place to start to see if it is true in this case.

We close this paper with a discussion of the local cohomology modules of a prime ideal in a complete regular local ring (R, m) of dimension d containing a algebraically closed residue field of characteristic 0 for $d \leq 5$.

In case $p = m$, everything is very well known. All the local cohomology vanishes except at d. If the $ht(p) = 1$, then p is principal and $cd(p, R) = 1$. The other extreme is when $\dim(R/p) = 1$. In this case, Hartshorne showed that $cd(p, R) = d - 1 = ht(p)$ (recall in general that $H_I^j(R) = 0$ for $j < grade(I) = ht(I)$ and is nonzero if $j = ht(I)$). If $\dim(R/p) = 2$, then (2.1) shows that $cd(p, R) = d - 2$. Putting these remarks together, we see that if $\dim(R) \leq 4$, then $cd(p, R) = ht(p)$ for all primes in R. The first interesting case is if $ht(p) = 2$ and $d = 5$. In this case (2.3) shows that $cd(p, R) \leq 3$. Let H be the third local cohomology with

support in p. Then $H_q = 0$ for all q with $ht(q) \leq 3$ by using
the Hartshorne-Lichtenbaum result mentioned above. What
about H_q if $ht(q) = 4$? Since R_q is not complete, we may lose
the primeness of p after completion and not be able to apply
(2.1)—it is even possible that the punctured spectrum   mod p
in the completion may not be connected. Thus it is possible
that $H_q \neq 0$ for infinitely many primes q of height 4. If there
were infinitely many, this would provide a counterexample to
the finiteness of associated primes of local cohomology. Essentially it comes down to this question: how many primes q
of height 4 have the property that the punctured spectrum of
$(R_q)\hat{\,}/p(R_q)\hat{\,}$ is disconnected? Besides the problem with the
completion there is also another problem. The residue field
$k(q)$ is no longer separably closed.

If it were true that $H_q = 0$ for all height 4 ideals (for
example in case that R/p is normal, or that R/p has an isolated singularity) then the support of H would be $\{m\}$. Since
$cd(p, R) \leq 3$, (4.1) applies to show that H is at least Artinian.

However Raghavan has pointed out that the last nonvanishing local cohomology module is never finitely generated, so H
could not be of finite length without being 0.

We close this paper with the observation that many of the
theorems listed in this paper are known only in the case the ring
contains a field. I feel sure all of them remain true in mixed
characteristic, and might not even be hard to show given a
concerted effort.

Bibliography

[B]   M. BRODMANN, *Einige Ergebnisse aus der lokalen Koho-mologietheorie und Ihre Anwendung*, Osnabrucker Schriften zur Mathematik, 5 (1983).

[BH] M. BRODMANN AND C. HUNEKE, *A quick proof of the Hartshorne-Lichtenbaum vanishing theorem*, preprint (1990).

[BR] M. BRODMANN AND J. RUNG, *Local cohomology and the connectedness dimension in algebraic varieties*, Comment. Math. Helvetici, 61 (1986), pp. 481–490.

[CS] F. W. CALL AND R. SHARP, *A short proof of the local Lichtenbaum-Hartshorne theorem on the vanishing of local cohomology*, Bull. London Math. Soc., 18 (1986), pp. 261–264.

[F1] G. FALTINGS, *Uber lokale Kohomologiegruppen hoher Ordnung*, J. fur d. reine u. angewandte Math., 313 (1980), pp. 43–51.

[F2] ————, *Uber die Annulatoren lokaler Kohomologiegruppen*, Arch. Math., 30 (1978), pp. 473–476.

[F3] ————, *Some theorems about Formal Functions*, Publ. RIMS Kyoto Univ., 16 (1980), pp. 721–737.

[G1] A. GROTHENDIECK, *Local Cohomology*, notes by R. Hartshorne, Lecture Notes in Math., 862, Springer-Verlag, 1966.

[G2] ————, *Cohomologie locale des faisceaux et theoremes de Lefshetz locaux et globaux (SGA 2)*, Amsterdam: North Holland Publ. Co. (1969).

[H1] R. HARTSHORNE, *Cohomological dimension of algebraic varieties*, Annals of Math., 88 (1968), pp. 403–450.

[H2] ————, *Affine duality and cofiniteness*, Inventiones Math., 9 (1970), pp. 145–164.

[HS] R. HARTSHORNE AND R. SPEISER, *Local cohomological dimension in characteristic p*, Annals of Math., 105 (1977), pp. 45–79.

[HuK] C. HUNEKE AND J. KOH, *Cofiniteness and vanishing of local cohomology*, preprint (1990).

[HuL] C. HUNEKE AND G. LYUBEZNIK, *On the vanishing of local cohomology*, Inventiones Math., 102 (1990), pp. 73–93.

[HuS] C. HUNEKE AND R. SHARP, *in preparation*.

[Ly] G. LYUBEZNIK, *Some Algebraic Sets of High Local Cohomological Dimension in Projective Space*, Proc. Amer. Math. Soc., 95 (1985), pp. 9–10.

[O] A. OGUS, *Local cohomological dimension of algebraic varieties*, Annals of Math., 98 (1973), pp. 327–365.

[PS] C. PESKINE AND L. SZPIRO, *Dimension Projective Finie et Cohomologie Locale*, I.H.E.S., 42 (1973), pp. 323–365.

[R] K. N. RAGHAVAN, *developing thesis*, Purdue Univ. (1991).

[S1] R. SPEISER, *Cohomological dimension of non-complete hypersurfaces*, Invent. Math., 21 (1973), pp. 143–150.

[S2] ———, *Cohomological dimension and abelian varieties*, Amer. J. Math., 95 (1973), pp. 1–34.

[S3] ———, *Projective varieties of low codimension in characteristic p > 0*, Trans. Amer. Math. Soc., 240 (1978), pp. 329–343.

Recent Work on Cremona Transformations
Sheldon Katz
Oklahoma State University
Stillwater, OK 74078

This survey article contains a very biased treatment of work done in the area of Cremona transformations during the past decade or so. As in classical times, this field is still rich in special cases with pretty descriptions, and divides into very different topics— no general theory has yet emerged. I have merely chosen the subjects that are most interesting to me. Most of these topics will be treated very briefly. A little more time will be taken up by a discussion of *special* Cremona transformations, i.e. those whose base scheme Y is smooth and irreducible. The connection to the theme of this conference is that certain rational mappings can be demonstrated to be Cremona transformations by using the syzygy matrix of the forms defining the mapping. This idea is due to Schreyer.

I have refrained from considering the more general subject of birational mappings between arbitrary varieties, especially in dimension 3. One instance of this is recent work extending the Iskovskikh-Manin approach to non-rationality of varieties (i.e. showing that there are few birational automorphisms). I will settle for merely listing two references here, the original Iskovskikh-Manin paper [IM] and its extension by Pukhlikov [P]. Another interesting topic not considered here is the work on Hanamura [H1,H2] which constructs the *scheme* $\mathrm{Bir}(X)$ of birational automorphisms of X, and shows that it behaves nicely if X is not uniruled. This uses some techniques related to the minimal model program (MMP). To my knowledge, the MMP has never been applied to the study of Cremona transformations.

I. Generalities

Definition. A *Cremona transformation* is a birational mapping from $\mathbf{P}^r$ to $\mathbf{P}^r$. The set of all Cremona transformations of $\mathbf{P}^r$ forms a group, denoted Cr_r.

Let $\Phi = (f_0, \cdots, f_r) : \mathbf{P}^r -- \to \mathbf{P}^r$ denote a Cremona transformation, with f_i forms of degree n on $\mathbf{P}^r$. The *base locus* X of Φ is the scheme of zeros of the f_i. We can blow up X (or more conveniently, perform a sequence of blow ups with smooth centers) to get a variety $\widetilde{\mathbf{P}}^r$ and a morphism

109

$\widetilde{\Phi} : \widetilde{\mathbf{P}}^r \to \mathbf{P}^r$ such that the following diagram commutes.

$$E \subset \widetilde{\mathbf{P}}^r$$

$$\downarrow \qquad \downarrow \pi \qquad \searrow \widetilde{\Phi}$$

$$X \subseteq \mathbf{P}^r \quad -\,-\to \quad \mathbf{P}^r$$

Here, $E = \cup E_i$ denotes the exceptional divisor, with irreducible components E_i. Let H denote the hyperplane class on $\mathbf{P}^r$. Then $\widetilde{\Phi}^* H = n\pi^* H - \Sigma m_i E_i$ for some $m_i \geq 0$.

Lemma 1 [CK1] $\dim |n\pi^* H - \Sigma m_i E_i| = r$, *i.e.* $\widetilde{\Phi}$ *is defined by a complete linear series.*

Proof: If the conclusion of the lemma were false, then Φ would factor through a projection $\mathbf{P}^{r'} \to \mathbf{P}^r$, with $r' > r$. The image of the factorization map must have degree 1, hence be linear. But mappings defined by complete linear systems are nondegenerate, a contradiction.   Q.E.D.

The moral is that alternatively, one can describe Cremona transformations by specifying a base locus X, degree n, and multiplicities m_i. The geometry of X then translates to information about the structure of Φ. For instance, if $C \subseteq \mathbf{P}^r$ is a curve whose proper transform $\widetilde{C} \subseteq \widetilde{\mathbf{P}}^r$ satisfies $\widetilde{C} \cdot (n\pi^* H - \Sigma m_i E_i) = 0$, then $\Phi(C)$ is a point. The nicest case is when $\widetilde{\Phi}$ is a blow down mapping (it is not true in dimension ≥ 3 that every birational mapping is the composition of blow ups and blow downs; counterexamples appear in [H], [O], and [C]). Two elementary examples come to mind.

1. The quadratic transformation $\Phi = (x_1 x_2, \ x_2 x_0, \ x_0 x_1) : \mathbf{P}^2 - - \to \mathbf{P}^2$.

   Here $X = \{p_1, \ p_2, \ p_3\} = \{(1,0,0), \ (0,1,0), \ (0,0,1)\}$. and $\widetilde{\Phi}^* H = 2\pi^* H - E_1 - E_2 - E_3$ if L_{ij} is the line joining p_i and p_j, then $\widetilde{L}_{ij} = \pi^* H - E_i - E_j$. Since $\widetilde{\Phi}^* H \cdot \widetilde{L}_{ij} = 0$, it follows that L_{ij} is blown down to a point. Hence Φ^{-1} is of the same type as Φ (in fact, $\Phi^2 = \mathrm{Id}$).

2. The second example is just as elementary, but less well known. Let $C \subseteq \mathbf{P}^3$ be an irreducible plane conic curve, with p a point of $\mathbf{P}^3$ not contained in the supporting plane P of C. The linear system of quadrics containing p and C define a Cremona transformation Φ. Let $L \subseteq \mathbf{P}^3$ be a line joining p to a point of C, or a line contained

in P. Then $\widetilde{\Phi}^*(H) \cdot \widetilde{L} = 0$. Hence Φ blows up a point and a conic, while blowing down P to a point, and the cone over C with vertex p gets blown down to a conic. Thus Φ^{-1} is of the same type as Φ. As an example, take $\Phi = (x_1 x_3 - x_2^2, \ x_0 x_1, \ x_0 x_2, \ x_0 x_3)$. Here $p = (1, 0, 0, 0)$, and C is defined by $x_0 = x_1 x_3 - x_2^2 = 0$.

II. Special Cremona Transformations

In a series of papers ([K], [CK1], [CK2]), Crauder and Katz have initiated the classification of Cremona transformations with smooth, irreducible base locus X. In the notation of §1, $\widetilde{\Phi}$ is given by the linear system $n\pi^* H - mE$. Let $L \subseteq \mathbf{P}^r$ denote a general line. Then we have

$$(n\pi^* H - mE)^r = 1, \text{ genus } \widetilde{\Phi}^{-1}(L) = 0$$

The genus of $\widetilde{\Phi}^{-1}(L)$ can be calculated by the adjunction formula. This yields a collection of Diophantine equations relating n, m, and the invariants of X. In low dimension, these equations can be solved. Some of the solutions are extraneous, in the sense that it can be shown by geometric reasoning that no X exists with the desired invariants. In the remaining cases, X, n, and m can be constructed, and usually even classified completely. Often, the most difficult part of the problem is to show that a rational mapping constructed from X, n, and m is indeed a Cremona transformation. A new technique for this part – the use of syzygies – has been developed in [HKS]. This technique will be discussed later. The other results in [HKS], together with those of [CK1] and [CK2] may be summarized as follows:

Theorem 1

I. *A Cremona transformation of* $\mathbf{P}^r$ *which becomes a morphism after the blow-up of* C, *a smooth curve of degree* d *and genus* g, *is given by the complete linear series of* n-*tics containing* C *where*

    i. $r = 3$, $d = 6$, $g = 3$ *and* $n = 3$ *or*

    ii. $r = 4$, $d = 5$, $g = 1$ *and* $n = 2$.

Conversely, if C *is a general curve of degree 6 and genus 3 in* $\mathbf{P}^3$ *or a nondegenerate smooth curve of degree 5 and genus 1 in* $\mathbf{P}^4$, *then the linear series above produce Cremona transformations.*

II. *If* Φ *is a Cremona transformation of* $\mathbf{P}^r$ *which becomes a morphism after the blow-up of* S, *a smooth surface, then* Φ *is given by the complete linear series of* n-*tics containing* S *where*

A. $r = 4$, $n = 3$ and S is a quintic elliptic scroll where $S = \mathbf{P}_C(\mathcal{E})$ with $e(\mathcal{E}) = -1$,

B. $r = 4$, $n = 4$ and S is a degree 10 determinantal surface given by the vanishing of the 4×4 minors of a 4×5 matrix of linear forms.

C. $r = 5$, $n = 2$ and S is the Veronese surface,

D_7. $r = 6$, $n = 2$ and S is a septic elliptic scroll where $S = \mathbf{P}_C(\mathcal{E})$ with $e(\mathcal{E}) = -1$, or

D_8. $r = 6$, $n = 2$ and S is $\mathbf{P}^2$ blown up at eight points and embedded in $\mathbf{P}^6$ as an octic surface by quartic curves passing simply through all eight points.

Conversely, in case A-D_7, any such smooth S gives a Cremona transformation. That of A is inverse to the quintic elliptic curve transformation of $\mathbf{P}^4$ given in Theorem I. ii, while B and C are self-dual. In case D_8, if no 4 points are colinear, and no 7 lie on a conic, then the resulting morphism Φ is a Cremona transformation.

III. *If Φ is a Cremona transformation of $\mathbf{P}^r$ which becomes a morphism after the blow up of X, a smooth threefold, then Φ is given by the complete linear series of n-tics containing X where*

A. $r = 5$ and X is given by the 5×5 minors of a 5×6 matrix of linear forms,

B. $r = 6$ $n = 3$, or

C. $r = 8$ $n = 2$

<u>Remarks</u>:

1. Cases II D_7 and II D_8 had been previously investigated by Semple and Tyrell [ST1], [ST2] by different methods.

2. In case IIIB, if X is the variety defined by the Pfaffians of the principal 6×6 submatrices of a 7×7 skew-symmetric matrix, then these 7 Pfaffians define a Cremona transformation.

3. Let Y be Fano threefold of index 1 and degree 14, so that $Y = G(2,6) \cap \mathbf{P}^9 \subset \mathbf{P}^{14}$. Let $X \subseteq \mathbf{P}^8$ be the projection of Y from a point on Y. If X is smooth, then the quadrics containing X define a Cremona transformation, giving an example of case IIIC.

4. In [CK2], the consequences of Hartshorne's conjecture on complete intersections were investigated. It turns out that in this case, $m = 1$,

which greatly simplifies the classification. There are many examples of possible invariants of varieties X which can only be eliminated by Hartshorne's conjecture. So one can fantasize that constructive methods for finding Cremona transformations with given invariants can be developed (some methods are given in [ESB] and [HKS]), leading to a counterexample to Hartshorne's conjecture.

Problem: Complete the classification of Cremona transformations whose base locus is a smooth threefold.

Special Cremona transformations have also been considered by Ein and Shepherd-Barron in [ESB]. They consider the following

Definition. A representation V of the semisimple group G is *regular prehomogeneous* if there is an invariant $P \in k[V]$ such that $\mathbf{P}(V^*) - (P = 0)$ is homogeneous under G, and the determinant of the Hessian of P is not identically zero.

Theorem 2 *If $V = \mathbf{C}^n$ is an irreducible regular prehomogeneous representation of the semisimple group G, then there is a coordinate system $(x_1, \ldots, x_n)$ on V such that if P is the unique irreducible invariant, then either*

(i) the rational map $T : \mathbf{P}^{n-1} \to \mathbf{P}^{n-1}$ given by

$$\left(\frac{\partial P}{\partial x_1}, \ldots, \frac{\partial P}{\partial x_n} \right)$$

is a Cremona transformation with $T^2 = 1$, or

(ii) $n = 2v$ is even and the map T given by

$$\left(\frac{\partial P}{\partial x_{v+1}}, \ldots, \frac{\partial P}{\partial x_{2v}}, -\frac{\partial P}{\partial x_1}, \ldots, -\frac{\partial P}{\partial x_v} \right)$$

is a Cremona transformation with $T^2 = 1$.

Recall that a smooth irreducible non-degenerate variety $X \subseteq \mathbf{P}^r$ is called a *Severi variety* if $\dim X = \frac{2r-4}{3}$ and its secant variety is not all of $\mathbf{P}^r$. There are only 4 Severi varieties, the Veronese in $\mathbf{P}^5$, the Segre embedding of $\mathbf{P}^2 \times \mathbf{P}^2$ in $\mathbf{P}^8$, the Plücker embedding of $G(2,6)$ in $\mathbf{P}^{14}$, and the E_6 variety in $\mathbf{P}^{26}$.

Theorem 3 *Let Φ be a Cremona transformation with smooth base locus X. Then Φ and Φ^{-1} are defined by quadrics if and only if X is a Severi variety.*

<u>Proof (sketch):</u> The Diophantine equations mentioned above are easily solved to show that X is a Severi variety. On the other hand, if X is a Severi variety, the secant variety is a cubic, defined by a polynomial P corresponding to a regular prehomogeneous vector space structure, and the partial derivatives of P cut out X scheme theoretically (since $X = Sing(Sec(x))$). Q.E.D.

Note that if Φ is defined by forms of degree n, then Φ blows down the n-secant variety of the base locus (the variety swept out by the n-secant lines of the base locus).

<u>Problem (Crauder):</u> Develop a theory of varieties with degenerate trisecant varieties analogous to the theory of Severi varieties. Relate this to the classification of Cremona transformations Φ such that Φ and Φ^{-1} are defined by cubics.

Ein and Shepherd-Barron have also proven the following.

Theorem 4 *Let $\Phi : \mathbf{P}^r \mathrel{-\,-\,\longrightarrow} \mathbf{P}^r$ be a Cremona transformation with smooth base locus X, codim $X = 2$. Then Φ is given by the complete linear series of n-tics containing X where*

- A. *$n = r = 3$, 4, or 5, and X is defined by the maximal minors of an $(n \times n+1)$ matrix of linear forms, or*

- B. *$r = 4$, $n = 3$, and X is a quintic elliptic scroll.*

The final topic considered on special Cremona transformations is Schreyer's insight that syzygies can be used to detect Cremona transformations [HKS].

Consider a map $\Phi = (f_0, \ldots, f_r) : \mathbf{P}^r \mathrel{-\,-\,\longrightarrow} \mathbf{P}^r$ as before, with base locus X and resolution $\widetilde{\Phi} : \widetilde{\mathbf{P}}^r \longrightarrow \mathbf{P}^r$. Let $Q = (Q_{ij})(0 \le i \le r)$ be any matrix of syzygies, i.e. $\sum_i f_i Q_{ij} = 0$ for all columns j. Let $a = (a_i) \in \mathbf{P}^r$ Let Q_a be the "generalized row" $(Q_a)_j = \sum_i a_i Q_{ij}$ of Q. Let $Z_a \subseteq \mathbf{P}^r$ be the scheme of zeros of the entries of Q_a. Let $f_a^\perp = \{\sum b_i f_i | \sum b_i a_i = 0\}$ be the pullback under Φ of the set of hyperplanes of $\mathbf{P}^r$ passing through a. Thus $V(f_a^\perp) = \pi(\tilde{\Phi}^{-1}(a)) \cup X$.

Theorem 5 *Suppose that for some $a \in \mathbf{P}^r$, Z_a is just $\{p\}$ for some $p \notin X$. Then the following are equivalent:*

1. *Rank $Q(p) = r$*

2. *$p \in V(f_a^\perp)$*

3. *Φ is dominant*

Furthermore, if any of these conditions hold, then Φ is a Cremona transformation.

Example. Here is how it was shown in [HKS] that the quadrics through any elliptic scroll of degree 7 and invariant $e = -1$ in $\mathbf{P}^6$ define a Cremona transformation. First it is shown by geometry and the Schrödinger representation of the Heisenberg group that any such scroll is projectively equivalent to an "H_7 equivariantly embedded elliptic scroll in $\mathbf{P}^6$." This means the following. Let $(x_0, \ldots, x_6)$ be homogeneous coordinates on $\mathbf{P}^6$, with the subscripts understood to be taken mod 7. Let $\sigma \in PGL_7$ be the automorphism inducing $\sigma(x_i) = x_{i+1}$. Then the H_7 equivariantly embedded elliptic scrolls are precisely the varieties cut out by the 7 quadrics $\{\sigma^i Q\}$, where

$$Q = -a_0 a_1 a_2 x_0^2 + a_0^2 a_1 x_1 x_6 - a_1^2 a_2 x_2 x_5 + a_2^2 a_0 x_3 x_4$$

where (a_0, a_1, a_2) satisfy $a_0^3 a_1 - a_1^3 a_2 - a_2^3 a_0 = 0$, $a_0 a_1 a_2 \neq 0$.

The quadrics $\sigma^i Q$ have the following matrix of syzygies, where the order of the quadrics is shown to the left.

$$
\begin{array}{c}
Q \\
\sigma Q \\
\sigma^2 Q \\
\sigma^3 Q \\
\sigma^4 Q \\
\sigma^5 Q \\
\sigma^6 Q
\end{array}
\left(
\begin{array}{ccccccc}
0 & -a_1 x_2 & a_2 x_4 & a_0 x_6 & -a_0 x_1 & -a_2 x_3 & a_1 x_5 \\
-a_1 x_5 & a_2 x_0 & a_0 x_2 & -a_0 x_4 & -a_2 x_6 & a_1 x_1 & 0 \\
a_2 x_3 & a_0 x_5 & -a_0 x_0 & -a_2 x_2 & a_1 x_4 & 0 & -a_1 x_1 \\
a_0 x_1 & -a_0 x_3 & -a_2 x_5 & a_1 x_0 & 0 & -a_1 x_4 & a_2 x_6 \\
-a_0 x_6 & -a_2 x_1 & a_1 x_3 & 0 & -a_1 x_0 & a_2 x_2 & a_0 x_4 \\
-a_2 x_4 & a_1 x_6 & 0 & -a_1 x_3 & a_2 x_5 & a_0 x_0 & -a_0 x_2 \\
a_1 x_2 & 0 & -a_1 x_6 & a_2 x_1 & a_0 x_3 & -a_0 x_5 & -a_2 x_0
\end{array}
\right)
$$

Apply theorem 5 to $a = (1, 0, \ldots, 0)$. Then $Z_a = \{(1, 0, \ldots, 0)\}$, and $Q(1, 0, \ldots, 0)$ has rank 6. Hence Φ is a Cremona transformation for the Heisenberg invariant scrolls, hence for all scrolls of the type considered.

For certain types of varietes, their equations can be reproduced from their syzygy matrices. Determinantal varieties are the most obvious example of this. This observation can be used to construct varieties with "many" linear syzygies ([HKS]). Theorem 5 now can be used to conclude that they give rise to Cremona transformations, if the syzygy matrices are properly chosen.

Problem: Find more families of Cremona transformations constructed from a linear syzygy matrix.

III. The Cremona Group Cr_r.

An old problem is to find generators and relations for Cr_r. The answer is known only for $r = 2$.

Let V be any rational variety. A rational mapping $V --- \rightarrow \mathbf{P}^r$ induces an isomorphism $Bir(V) \simeq Cr_r$ of the Cremona group with the group of birational automorphisms of V. Here, two answers for $r = 2$ are reviewed.

First, it was known classically that Cr_2 is generated by the group PGL_3 of automorphisms of $\mathbf{P}^2$, together with the quadratic transformations.

The relations have been studied by Gizatullin [G], who found that the relations are generated by relations of the form $g_1 g_2 g_3 = 1$. Both this fact, and the classical fact about quadratic transformations and automorphisms generating Cr_2 follow from an argument in combinatorial group theory applied to a graph which encodes the data of all possible birational maps of a rational variety to $\mathbf{P}^2$, together with the relations between these maps arising from composition with a quadratic transformation.

Iskovskikh [I] has found a different solution to the problem. Let F_0 denote the rational ruled surface $\mathbf{P}^1 \times \mathbf{P}^1$. He first proves that $Bir(F_0)$ is generated by the "switch involution" $\tau(x, y) = (y, x)$, the biregular automorphisms $PGL(2) \times PGL(2)$, and the involution $e(((u_0 : u_1), (v_0 : v_1))) = ((u_0 : u_1), (u_0 v_1, u_1 v_0))$. He then shows that the relations are somewhat more explicit than the relations between the quadratic transformations mentioned above.

Problem: Find generators and relations for Cr_3.

In [U1], [U2], [U3], and [U4], Umemura has found the maximal connected algebraic subgroups of Cr_3.

Let X be a rational threefold, and G an algebraic group acting effectively on X. A birational map $f : X -- \rightarrow \mathbf{P}^3$ naturally determines an embedding $G \hookrightarrow Cr_3$; hence the action of G on X determines a conjugacy class of a subgroup of Cr_3, whose members are isomorphic to G. Both the action of G and the resulting conjugacy class in Cr_3 will be referred to as algebraic operations. The conjugacy classes of maximal subgroups will be described as algebraic operations.

Before stating the result, some names are needed for certain rational varieties.

Let $F'_m = Spec\left(\bigoplus_{k\geq 0}\mathcal{O}_{\mathbf{P}^1}(-km)\right)$ be the total space of the line bundle of degree m over $\mathbf{P}^1$.

Let $J'_m = Spec\left(\bigoplus_{k\geq 0}\mathcal{O}_{\mathbf{P}^2}(km)\right)$ be the total space of the line bundle of degree m over $\mathbf{P}^2$.

Let $L'_{m,n} = Spec\left(\bigoplus_{k\geq 0}\mathcal{O}_{\mathbf{P}^1\times\mathbf{P}^1}(-km,-kn)\right)$ be the total space of the line bundle of bidegree (m,n) over $\mathbf{P}^1\times\mathbf{P}^1$.

Let $F'_{m,n} = Spec\left(Sym(\mathcal{O}_{\mathbf{P}^1}(-n)\bigoplus\mathcal{O}_{\mathbf{P}^1}(-n)\right)$ be the total space of the vector bundle $\mathcal{O}_{\mathbf{P}^1}(m)\bigoplus\mathcal{O}_{\mathbf{P}^1}(n)$ over $\mathbf{P}^1$.

E''_m denotes a particular $\mathbf{A}^1$ bundle over F'_m, depending on l. See [U4] for details.

Theorem 6 (I) *Let G be a connected algebraic group in Cr_3. Then G is contained in the conjugacy class of one of the following algebraic operations:*

> (P1) $(PGL_4, \mathbf{P}_3)$,
> (P2) $(PSO_5, \; quadric \subset \mathbf{P}_4)$.
> (E1) $(PGL_2, PGL_2/\Gamma)$, Γ *is an octahedral subgroup of* PGL_2.
> (E2) $(PGL_2, PGL_2/\Gamma)$, Γ *is an icosahedral subgroup of* PGL_2.
> (J1) $(PGL_3 \times PGL_2, \mathbf{P}^2 \times \mathbf{P}^1)$.
> (J2) $(PGL_2 \times PGL_2 \times PGL_2, \mathbf{P}^1 \times \mathbf{P}^1 \times \mathbf{P}^1)$.
> (J3) $(PGL_2 \times Aut^0 F'_m, \mathbf{P}^1 \times F'_m)$ *where m is an integer ≥ 2.*
> (J4) $(PGL_3, PGL_3/B)$ *where B is a Borel subgroup of* PGL_3.
> (J5) $(PGL_2, PGL_2/D_{2n})$ *where n is an integer ≥ 4.*
> (J6) $(G, G/H_{m,n})$ *where* $G = \mathbf{G}_m \times SL_2 \times SL_2$

$$H_{m,n} = \left\{\left(t_1^m t_2^n, \begin{pmatrix} t_1 & x \\ 0 & t_1^{-1} \end{pmatrix}, \begin{pmatrix} t_2 & y \\ 0 & t_2^{-1} \end{pmatrix}\right) \middle| t_1, t_2 \in k^*, x, y \in k\right\}$$

> *and m,n are integers with $m > 2, -2 > n$.*
> (J7) $(Aut^0 J'_m, J'_m)$ *where m is an integer $m \geq 2$.*
> (J8) $(Aut^0 L'_{m,n}, L'_{m,n})$ *where m,n are integers with $m \geq n \geq 1$.*
> (J9) $(Aut^0 F'_{m,n}, F'_{m,n})$ *where m,n are integers with $m > n \geq 2$.*
> (J10) $(Aut^0 F'_{m,m}, F'_{m,m})$ *where m is an integer ≥ 2.*

(J11) $(\mathrm{Aut}^0 E_m^{\prime l}, E_m^{\prime l})$ *where* l, m *are integers with* $m \geq 2,\ l \geq 2$ *or* $m = 1,\ 1 \geq 3.$

(J12) *Generically intransitive operation* (PGL_2, X_π) *with general orbits isomorphic to* $(PGL_2, PGL_2/\mathbf{G}_m)$, *where* $\pi : C_1 \to C_2$ *is an étale 2-covering of a rational curve* C_2 *with genus* $(C_1) \geq 1.$ *(These operations are effectively parametrized by the moduli space of nonsingular elliptic or hyperelliptic curves of genus* ≥ 1*).*

(II) *The (conjugacy classes of) algebraic subgroups of* Cr_3 *determined by the above operations* (P1), (P2), (E1), (E2), (J1), $\cdots$,(J12) *are maximal (conjugacy classes of) algebraic subgroups of* Cr_3.

References

[C]   B. Crauder. Birational morphisms of smooth threefolds collapsing three surfaces to a point. Duke Math. J. **48** (1981) 589–632.

[CK1]   B. Crauder and S. Katz. Cremona transformations with smooth irreducible fundamental locus. Amer. J. Math. **111** (1989) 289–309.

[CK2]   B. Crauder and S. Katz. Cremona transformations and Hartshorne's conjecture. Amer. J. Math. **113** (1991) 269–285.

[ESB]   L. Ein and N. Shepherd-Barron. Some special Cremona transformations. Amer. J. Math. **111** (1989) 783–800.

[G]   M.H. Gizatullin. Defining relations for the Cremona group of the plane. Math. USSR-Izv. **21** (1983) 211–268.

[H]   H. Hironaka. An example of a non-Kahlerian complex-analytic deformation of Kahlerian complex structures. Ann. Math. **75** (1962) 190–208.

[H1]   M. Hanamura. On the birational automorphism groups of algebraic varieties. Comp. Math. **63** (1987) 123–142.

[H2]   M. Hanamura. Structure of birational automorphism groups, I: non-uniruled varieties. Inv. Math. **93** (1988) 383–403.

[HKS]   K. Hulek, S. Katz, and F.-O. Schreyer. Cremona transformations and syzygies. Math. Z., to appear.

[I]   V.A. Iskovskikh. Proof of a theorem on relations in the two-dimensional Cremona group. Russian Math Surveys **40** (1985) 231–232.

[IM] V.A. Iskovskikh and Yu.I. Manin. Three-dimensional quartics and counterexamples to the Lüroth problem. Math. USSR Sbornic **15** (1971) 141–165.

[K] S. Katz. The cubo-cubic transformation of P^3 is very special. Math. Z. **195** (1987) 255–257.

[O] T. Oda. Convex Bodies and Algebraic Geometry. Introduction to the Theory of Toric Varieties. Springer-Verlag. Berlin-Heidelberg-New York 1988.

[P] A.V. Pukhlikov. Birational isomorphisms of four-dimensional quintics. Inv. Math. **87** (1987) 303–329.

[ST1] J.G. Semple and J.A. Tyrrell. The Cremona transformation of S_6 by quadrics through a normal elliptic septimic scroll $^1R^7$. Mathematika **16** (1969) 88–97.

[ST2] J.G. Semple and J.A. Tyrrell. The $T_{2,4}$ of S_6 defined by a rational surface $^3F^8$. Proc. London Math. Soc. (3) **20** (1970) 205–221.

[U1] H. Umemura. Sur les sous-groupes algébriques primitifs du groupe de Cremona à trois variables. Nagoya Math. J. **79** (1980) 47-67.

[U2] H. Umemura. Maximal algebraic subgroups of the Cremona group. Nagoya Math. J. **87** (1982) 59–78.

[U3] H. Umemura. On the maximal connected algebraic subgroups of the Cremona group I. Nagoya Math. J. **88** (1982) 213–246.

[U4] H. Umemura. On the maximal connected algebraic subgroups of the Cremona group II. In: Algebraic groups and related topics (Kyoto/Nagoya, 1983), 349–436. Adv. Stud. Pure Math. 6 North-Holland, Amsterdam-New York 1985.

The Homological Conjectures

Paul C. Roberts

In 1975 Hochster [12] published a monograph in which he discussed a group of conjectures, often referred to as the "Homological Conjectures", since they are all related in one way or another to homological algebra, and, in particular, to properties of modules of finite projective dimension. Since that time there has been progress in several directions; some of the conjectures have been proven, some shown to be false, and many new conjectures have been added to the list. We do not intend to give an exhaustive list here of all the developments in the field, but rather to summarize the present state of research on these conjectures, discussing which of these questions are still open and the simplest cases for which they are not known.

We begin by reproducing Hochster's diagram:

The Homological Conjectures (Hochster 1975)

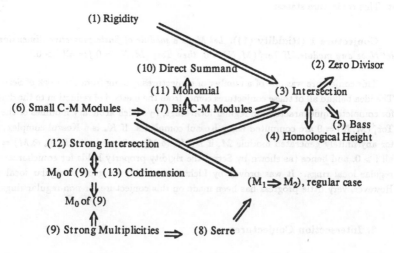

For complete statements of all of these conjectures we refer to Hochster [12]. We will discuss most of them here, and will refer to their number in the above diagram by a number in parentheses. We collect these conjectures into groups of similar problems. The outline is as follows:

121

1. Rigidity (1).

2. Intersection (2), (3), (4), (5).

3. Cohen-Macaulay modules (6), (7).

4. Monomial (10), (11).

5. Multiplicities (8), (9).

6. Codimension (12), (13).

1. Rigidity.

We first discuss the Rigidity Conjecture, partly because it is at the top of the diagram, and partly because so little progress has been made that it will not take long to deal with it. This conjecture states:

Conjecture 1 (Rigidity (1)). *Let M be a module of finite projective dimension and let N be any module. If $Tor_1^A(M,N) = 0$, then $Tor_i^A(M,N) = 0$ for all $i > 0$.*

This conjecture was one of a number of conjectures coming from the work of Serre [28]. The idea behind all of these conjectures comes from the method of reduction to the diagonal for complete equicharacteristic regular local rings, by which arbitrary modules of the form $Tor_1^A(M,N) = 0$ are computed using Koszul complexes. If $K_\bullet$ is a Koszul complex, then, for any finitely generated module M, if $H_1(K_\bullet \otimes M) = 0$, we have $H_i(K_\bullet \otimes M) = 0$ for all $i > 0$, and hence (as shown by Serre) the rigidity property holds for equicharacteristic regular local rings. It was proven by Lichtenbaum [17] for general regular local rings. However, very little progress has been made on this conjecture for non-regular rings.

2. Intersection Conjectures.

These conjectures ((2) through (5)) derive from the Intersection Theorem of Peskine and Szpiro [18],[19] and its consequences, and are now known in general (Peskine-Szpiro [19] for rings of positive characteristic or essentially of finite type over a field, Hochster [11] for the general equicharacteristic case, Roberts [25],[26] for the mixed characteristic case). We state a newer version of the basic theorem:

Theorem (New Intersection Theorem: A newer version of (3)). *Let* $F_\bullet =$

$$0 \to F_k \to \ldots \to F_0 \to 0$$

be a non-exact complex of free A-modules with homology of finite length. Then

$$k \geq \dim(A).$$

There are various refinements of this theorem which are still conjectural. In one direction, this theorem can be interpreted as a lower bound on the rank of the dth free module in a complex $F_\bullet$ satisfying the hypotheses of the thoerem and with $H_0(F_\bullet) \neq 0$ (that is, $F_d \neq 0$, so the lower bound on the rank is 1). There are other conjectural lower bounds for the ranks of the other free modules in such a complex which are open even for regular local rings; we refer to the article on Betti numbers by Charamboulos and Evans [3] in these proceedings for a discussion of these questions.

Another stronger version of the Intersection Theorem is:

Conjecture 2 (Improved New Intersection Conjecture). *Let* $F_\bullet =$

$$0 \to F_k \to \ldots \to F_0 \to 0$$

be a non-exact complex of free A-modules with homology of finite length except possibly for $H_0(F_\bullet)$. *Assume there is a minimal generator of* $H_0(F_\bullet)$ *annihilated by a power of the maximal ideal of A. Then*

$$k \geq \dim(A).$$

This conjecture was introduced by Evans and Griffith [6] as a lemma in the proof of their Syzygy Theorem. It is known for equicharacteristic rings, where it can be proven by reduction to positive characteristic and use of the Frobenius map. However, the technique of using the local Chern characters of Baum, Fulton and MacPherson [2], which was used in the proof of the New Intersection Theorem in mixed characteristic, does not appear to extend to the improved version, and this question is still open in mixed characteristic. We will return to this conjecture in the section on the Monomial Conjecture, since results of Hochster [13] and Dutta [4] have shown that it is equivalent to the Monomial Conjecture.

3. Cohen-Macaulay Modules.

Conjecture 3 (Small Cohen-Macaulay modules (6)). *Let A be a complete domain of dimension d. Then there exists a finitely generated A-module of depth d.*

Such a module can be constructed easily in dimension 2 by taking the integral closure of A. This conjecture is open in dimension three and higher in every characteristic. There does not appear to be a natural way of constructing such modules, but, on the other hand, there does not seem to be a way of showing that no such modules exist over a given ring. We remark that since A is assumed to be a complete domain, we can find a regular local subring over which it is a finite module, and, since a small Cohen-Macaulay module would be free over R, the existence of a small Cohen-Macaulay module is equivalent to the existence of an R-algebra homomorphism from A to a ring of matrices over R. It is easy to construct examples where no such homomorphism exists in certain cases when A is a finite extension of a ring R which is not regular.

Conjecture 4 (Big Cohen-Macaulay modules (7)). *Let A be a local ring of dimension d, and let $x_1, \ldots x_d$ be a system of parameters for A. Then there exists an A-module M (not necessarily finitely generated) such that $(x_1, \ldots x_d)M \neq M$ and such that $x_1, \ldots x_d$ is a regular sequence on M.*

A module M satisfying the conclusion of this conjecture is called a big Cohen-Macaulay module. Big Cohen-Macaulay modules were introduced by Hochster [12], who showed that their existence implies many other conjectures, as shown in the diagram reproduced in the introduction. He proved that big Cohen-Macaulay modules do exist for equicharacteristic local rings. Griffith [10] has extended this and shown that (still for equicharacteristic rings) if A is a finite extension of a regular local ring R, then a non-zero A-module exists which is free over R. This conjecture is open in mixed characteristic for dimension 3 and higher, and it appears to be of roughly the same level of difficulty as the Monomial Conjecture, although it is somewhat stronger.

In a slightly different direction, one can define the notion of "Cohen-Macaulay complexes" in such a way that their existence has most of the same implications as the existence of big Cohen-Macaulay modules. The existence of Cohen-Macaulay complexes can be shown over the complex numbers by analytic methods (Roberts [22]).

Recently Hochster and Huneke [15] have shown that in positive characteristic, the integral closure of a domain in the algebraic closure of its quotient field is Cohen-Macaulay in this sense. This shows the existence of a big Cohen-Macaulay algebra (in the equicharacteristic case), and it also provides a natural construction of one for rings of positive characteristic. Using this construction many of the consequences of the existence of big Cohen-Macaulay modules can be strengthened or their proofs simplified.

4. The Monomial Conjecture.

The conjectures discussed here are all known in the equicharacteristic case and unknown in mixed characteristic.

Conjecture 5 (Monomial Conjecture (11)). *Let $x_1, \ldots, x_d$ be a system of parameters for A. Then $x_1^t x_2^t \ldots x_d^t$ is not in the ideal generated by $x_1^{t+1}, \ldots x_d^{t+1}$.*

The importance of this conjecture was shown by Hochster [13]. The crucial situation for this conjecture, as shown by Hochster, is when A is a finite extension of a complete regular local ring R and the system of parameters is a regular system of parameters for R. Furthermore, (and this is not so obvious) R can be assumed to be unramified and thus a power series ring over $\hat{Z}_p$. The (equivalent) Direct Summand Conjecture (10) states that R is a direct summand of A in this situation. This conjecture is trivially true when the degree of the extension is prime to p by using the trace map; apart from this, the strongest general result, due to Koh [16], states that the result is true for Galois extensions of degree p (and some extensions of degree p^2). It has been shown (Hochster[13], Dutta[4]) that the Monomial Conjecture is equivalent to the "Canonical Element" conjecture and to the Improved New Intersection Conjecture mentioned above. The Canonical Element conjecture can be stated in several ways (see Hochster [13]). We give one version here.

Conjecture 6 (Canonical Element Conjecture). *Let $x_1, \ldots x_d$ be a system of parameters for A. Let $K_\bullet$ be the Koszul complex on $x_1, \ldots x_d$, let $F_\bullet$ be a minimal free resolution of $A/(x_1, \ldots x_d)$, and let $\phi_\bullet : K_\bullet \to F_\bullet$ be a map of complexes lifting the identity map in degrees 0 and 1. Then $\phi_d(K_d) \not\subseteq m F_d$ (where m is the maximal ideal of A.)*

We describe one recent approach to the Monomial Conjecture in dimension 3, an approach which also shows the relationship with some of the other conjectures. Denote the system of parameters X, Y, Z (when the ring has mixed characteristic p, we will let Z denote p.) Let R be a regular subring of A such that X, Y, Z is a regular sequence of parameters for R. Let I be the ideal generated by $X^{t+1}, Y^{t+1}, Z^{t+1}, X^t Y^t Z^t$, and let $F_\bullet$ denote the free resolution of R/I over the regular subring, tensored with A. The Monomial Conjecture states that there is no element of the form $(a, b, c, 1)$ in the kernel of $F_1 \to F_0$.

Conjecture 7. *If A is a local ring of dimension d and*

$$0 \to F_d \to \ldots \to F_0 \to 0$$

is a complex of free modules with homology of finite length, then the module of cycles is integral over the module of boundaries in every degree > 0.

In the case under consideration, the element $(a, b, c, 1)$ could not possibly be integral over the boundaries, since the boundaries have entries in the maximal ideal. Evidence for this conjecture is that it holds in the equicharacteristic case via tight closure (see Hochster and Huneke [14]); it has also been proven by Rees [21] in the case of the Koszul complex on a system of parameters.

Examination of this question leads to the study of the algebra generated by the boundaries in the symmetric algebra on F_1. Let this symmetric algebra be $A[S, T, U, V]$, where the indeterminates correspond to the generators of the ideal I. Let D denote the derivation

$$X^{t+1} \frac{\partial}{\partial S} + Y^{t+1} \frac{\partial}{\partial T} + Z^{t+1} \frac{\partial}{\partial U} + X^t Y^t Z^t \frac{\partial}{\partial V}.$$

If there exists an extension of R over which an element of the kernel of D exists of the form $aS + bT + cU + V$ (this is what a counterexample to the Monomial conjecture looks like in this form), its norm would be in the kernel of the derivation D and have coefficients in R with V^n coefficient 1. Conversely, such an element of $R[S, T, U, V]$ will give rise to a counterexample if it has a linear factor over some extension of the quotient field of R. Since the existence of a linear factor in an extension field can be expressed by the vanishing of equations on the coefficients, this makes the whole problem one of the existence of solutions to equations (albeit complicated ones) in the regular ring $R[S, T, U, V]$ without reference to a specific extension.

Finally, we note that if there is an element $(a, b, c, 1)$ of F_1 which goes to zero in F_0, then the truncated complex

$$0 \to F_3 \to F_2 \to F_1 \to 0$$

is a counterexample to the Improved New Intersection Conjecture, since this element defines a minimal generator of the homology at F_1 which is annihilated by a power of the maximal ideal. Thus, since the Monomial Conjecture implies this conjecture, if there is a counterexample to the Improved New Intersection Conjecture (in dimension 3), then this particular complex must be a counterexample for some extension of R.

5. Multiplicity Conjectures.

These conjectures include the conjectures of Serre on intersection multiplicities defined by Euler characteristics over regular local rings and various generalizations of them. Like the Rigidity Conjecture, the multiplicity conjectures generalize properties of Koszul complexes. We first state them in a general form:

Conjecture 8 (Multiplicities ((8) for regular rings, (9) for arbitrary rings)). *Let M be a module of finite projective dimension, and let N be a module such that $M \otimes_A N$ is a module of finite length. Let $\chi(M, N) = \sum (-1)^i \text{length}(\text{Tor}_i^A(M, N))$. Then we have*

(M$_0$) $\dim(M) + \dim(N) \leq \dim(A)$.

(M$_1$) **(Vanishing)** *If $\dim(M) + \dim(N) < \dim(A)$, then $\chi(M, N) = 0$.*

(M$_2$) **(Positivity)** *If $\dim(M) + \dim(N) = \dim(A)$, then $\chi(M, N) > 0$.*

The first of these three statements was proven by Serre for regular rings. It is also true if M has projective dimension 1 by using the determinant and Krull's Principal ideal theorem. It is not known in general for non-regular rings in any characteristic.

The vanishing conjecture was proven by Serre [28] in the equicharacteristic regular case, and by Roberts [23] and Gillet-Soulé [8], [9] in the general regular case. In addition, whenever both modules have finite projective dimension and the Euler characteristic is closely related to local Chern characters (see Roberts [24]) vanishing is known; this includes complete intersections and any ring with singularity locus of dimension at most 1. We note that since M$_1$ is now known for regular rings, the diagonal arrow from small CM modules in the diagram in the introduction can end simply at M$_2$, so that the existence of small Cohen-Macaulay modules implies the Serre Positivity Conjecture.

The entire Multiplicity Conjecture has been proven for graded modules over graded rings by Peskine and Szpiro [20]. Conjectures M$_1$ and M$_2$ are also true for general local rings if N has dimension 1 (this was shown by Foxby [7]).

In the case in which only one module has finite projective dimension, M$_1$ and M$_2$ are false, as shown by the example of Dutta, Hochster, and McLaughlin [5]. They construct an example where $A = k[[X, Y, Z, W]]/(XY - ZW)$, where M is a module of finite length (its length is 15) and finite projective dimension, and where $\chi(M, N) = -1$ for $N = A/(X, Z)$. Thus vanishing, and hence also positivity, is false in this generality.

The positivity conjecture was proven for equicharacteristic regular rings by Serre. For mixed characteristic regular rings it is still open. We remark on one partial result in this case.

Let A be a regular local ring, and let P and Q be prime ideals such that $\dim(A/P) + \dim(A/Q) = \dim(A)$ and such that $P + Q$ is primary to the maximal ideal. The usual reduction, using the fact that all modules have finite projective dimension, shows that this case implies the general case when the ring is regular. If the subvarieties defined by P and Q are not tangent, then the intersection multiplicity is simply the product of the multiplicities of A/P and A/Q, which is positive (Tennison [29]). In general there is another term, which should itself be positive if the two subvarieties are tangent at the point.

In the non-regular case, it is seen from the example of Dutta, Hochster, and McLaughlin mentioned above that positivity is false if only one of the modules is required to have finite projective dimension. If we require both modules to have finite projective dimension, the question is open. On the other hand, an example (see Roberts [27]) shows that positivity is not so natural in this case. There exists a (non-regular) ring A of dimension 4 and two complexes $F_\bullet$ and $G_\bullet$ of free A modules with the following properties: first, the supports of the complexes have dimension two and intersect only at the closed point. Second, for every ideal P of height two, the Euler characteristic $\chi((F_\bullet)_P)$ is non-negative, and for one such P it is positive, and the same holds for $G_\bullet$. On the other hand, $\chi(F_\bullet \otimes G_\bullet) < 0$. We remark that for regular local rings, if positivity as stated in M_2 above holds, then no example with these properties can exist. However, there is no known example of the failure of the Positivity Conjecture in which $F_\bullet$ and $G_\bullet$ are resolutions of modules.

6. Grade (Codimension) Conjectures.

I would like to thank Hans-Bjørn Foxby for many useful suggestions in preparing this section.

The main conjecture, due to M. Auslander, states the following:

Conjecture 9 (Codimension (13)). *Let M be a module of finite projective dimension. Then $\mathrm{grade}(M) = \dim(A) - \dim(M)$.*

We recall that the grade (also sometimes called codimension, which gave the name to this conjecture) of a module M is the maximal length of a regular sequence contained in the annihilator of M. Equivalently, the grade of M is the smallest integer i such that $\mathrm{Ext}_A^i(M, A) \neq 0$. Thus if I is the annihilator of M, this conjecture states that there is a regular sequence of length $\dim(A) - \dim(M)$ in I. If M has finite length, this result states that A is Cohen-Macaulay, which is a special case of the New Intersection Theorem. In fact, more can be deduced from the Intersection Theorem: suppose there is a maximal regular sequence for A contained in the annihilator of M (so that after dividing by this sequence there is an element annihilated by the maximal ideal m of A). Let k be the dimension of M, let $y_1, \ldots y_k$ be a sequence of elements of m such that $M/(y_1, \ldots y_k)$ has finite length, and let $B = A/(y_1, \ldots y_k)$. Let r be grade of M; under our assumption, r is also the depth of A. By the Auslander-Buchsbaum equality the projective dimension of M is less than or equal to r. If $F_\bullet$ is a projective resolution of M, then tensoring $F_\bullet$ with B gives a complex satisfying the hypotheses of the New Intersection Theorem, so we have $r \geq \dim(B) = \dim(A) - k$. Thus we have $\mathrm{grade}(M) + \dim(M) \geq \dim(A)$, so the Grade Conjecture holds in this case.

Now let M be any module of finite projective dimension and let r denote the grade of M. Let $k = \dim(A) - \dim(M)$. Suppose that for each minimal prime ideal in the support of M we have $\mathrm{height}(P) \geq k$. If $x_1, \ldots x_r$ is a maximal regular sequence in the annihilator of M, we can find an associated prime ideal P of $A/(x_1, \ldots x_r)$ containing the annihilator of M; localizing at P, we are in the situation where $x_1, \ldots x_r$ is a maximal regular sequence for A_P, so, by the above discussion, we have $r = \dim(A_P) - \dim(M_P) \geq k$. Thus the Grade Conjecture holds.

It follows from this discussion that the Grade Conjecture is really a question on the height of minimal prime ideals in the support of a module of finite projective dimension. By completing, we can assume that the ring A is catenary. If A is also equidimensional, then the Grade Conjecture holds for A-modules of finite projective dimension.

Peskine and Szpiro [20] showed that if vanishing holds (M_1 above in the section on multiplicity conjectures), then the codimension conjecture holds; using this result, they proved that the Grade Conjecture holds for graded modules over graded rings. While M_1 does not hold in general, it does in codimensions 0 (this case is quite easy) and 1, and thus the codimension conjecture holds in these cases; it is sometimes possible to show that the multiplicity conjectures hold in a specific case, so that the Grade Conjecture also holds in that case. Putting these results together, one can deduce that the lowest unknown case is when A has dimension 4 and depth 3, and M has grade 2, dimension 1, and depth 0. The conjecture is unknown for cyclic modules in this case.

In case M is not only a module but and A-algebra (still of finite homological dimension), there is another version of the Grade Conjecture due to Avramov and Foxby [1]. For a local ring S, we define the *Cohen-Macaulay defect* (abbreviated cmd) of S as follows:

$$\mathrm{cmd}(S) = \dim(S) - \mathrm{depth}(S).$$

If we define the amplitude of a complex to be the difference between the degrees of the highest and lowest non-vanishing homology of the complex, then the Cohen-Macaulay defect of A is the amplitude of a dualizing complex for A. If $\phi : A \rightarrow S$ is a ring homomorphism such that S is a finite A-module, one can define the dualizing complex of the homomorphism ϕ to be $\mathrm{RHom}(S, A)$; we then define the Cohen-Macaulay defect of ϕ to be the amplitude of this complex.

Proposition. *The Grade Conjecture holds for all module finite algebras S over A of finite projective dimension if and only if for all such extensions we have the equality*

$$\mathrm{cmd}\phi = \mathrm{cmd}S - \mathrm{cmd}A.$$

In this form the conjecture can be generalized to non-finite extensions of finite flat dimension (see Avramov and Foxby [1]).

References

1. L. Avramov and H.-B. Foxby, *Locally Cohen-Macaulay homomorphisms*, in preparation.

2. Baum, P., Fulton, W., MacPherson, R. *Riemann-Roch for singular varieties*, Publ. Math. IHES **45** (1975) 101-145.

3. H. Charalambous and E. G. Evans, *Problems of Betti numbers of finite length modules*, these Proceedings.

4. S.P. Dutta, *On the canonical element conjecture*, Trans. Amer. Math. Soc. **299** (1987) 803-811.

5. S.P. Dutta, M. Hochster, and J.E. McLaughlin, *Modules of finite projective dimension with negative intersection multiplicities*, Invent. Math. **79** (1985) 253-291.

6. E.G. Evans, and P. Griffith, *The syzygy problem: a new proof and historical perspective*, Commutative Algebra (Durham 1981), London Math Soc. Lecture Note Series **72** (1982) 2-11.

7. H.-B. Foxby, *The MacRae invariant*, Commutative Algebra (Durham 1981), London Math Soc. Lecture Note Series **72** (1982) 121-128.

8. H. Gillet and C. Soulé, *K-théorie et nullité des multiplicités d'intersection*, C. R. Acad. Sc. Paris Série I no. 3, t. **300** (1985) 71-74.

9. H. Gillet and C. Soulé, *Intersection theory using Adams operations*, Invent. Math. **90** (1987) 243-277.

10. P. Griffith, *A representation theorem for complete local rings*, J. of Pure and Applied Algebra **7** (1976) 303-315.

11. M. Hochster, *The equicharacteristic case of some homological conjectures on local rings*, Bull. Amer. Math. Soc. **80** (1974) 683-686.

12. M. Hochster, *Topics in the homological theory of modules over commutative rings*, Regional Conference Series in Mathematics **24**, 1975.

13. M. Hochster, *Canonical elements in local cohomology modules and the direct summand conjecture*, Journal of Algebra, **84** (1983) 503-553.

14. M. Hochster and C. Huneke, *Tight closure, invariant theory, and the Briançcon-Skoda Theorem*, J. of the Amer. Math. Soc. **1** (1990) 31-116.

15. M. Hochster and C. Huneke, *Infinite Integral Extensions and big Cohen-Macaulay algebras*, preprint.

16. J. Koh, *Degree p extensions of an unramified regular local ring of mixed characteristic p*, J. of Algebra **99** (1986) 310-323.

17. S. Lichtenbaum, *On the vanishing of Tor in regular local rings*, Ill. J. of Math. **10** (1966) 220-226.

18. C. Peskine and L. Szpiro, *Sur la topologie des sous-schémas fermés d'un schéma localement noethérien, définis comme support d'un faisceau cohérent localement de dimension projective finie*, C. R. Acad. Sci. Paris, Sér.A **269** (1969) 49-51.

19. C. Peskine and L., Szpiro, *Dimension projective finie et cohomologie locale*, Publ. Math. IHES **42** (1973) 47-119.

20. C. Peskine and L. Szpiro, *Syzygies et Multiplicités*, C. R. Acad. Sci. Paris, Sér.A **278** (1974) 1421-1424.

21. D. Rees, *Reduction of Modules*, Math. Proc. Camb. Philos. Soc. **101** (1987) 431-449.

22. P. Roberts, *Cohen-Macaulay complexes and an analytic proof of the New Intersection Conjecture*, J. of Algebra **66** (1980) 220-225.

23. P. Roberts, *The vanishing of intersection multiplicities of perfect complexes*, Bull. Amer. Math. Soc. **13** (1985) 127-130.

24. P. Roberts, *Local Chern characters and intersection multiplicities*, Proc. Sympos. Pure Math. **46** 2, Amer. Math. Soc., Providence, R. I. (1987) 389-400.

25. P. Roberts, *Le théorème d'intersection*, C. R. Acad. Sc. Paris, Sér.I no. 7, t. **304** (1987) 177-180.

26. P. Roberts, *Intersection Theorems*, Commutative Algebra, Proceedings of an MSRI Microprogram, Springer-Verlag, (1989) 417-436.

27. P. Roberts, *Negative intersection multiplicities on singular varieties*, to appear in the proceedings of the Zeuthen Conference, Copenhagen 1989.

28. J.-P. Serre, Algèbre Locale - Multiplicités. Lecture Notes in Mathematics vol. 11, Springer-Verlag, New York, Berlin, Heidelberg, 1961.

29. B. R. Tennison, *Intersection multiplicities and tangent cones*, Math. Proc. Camb. Philos. Soc. **85** (1979) 33-42.

Remarks on Residual Intersections

Bernd Ulrich*

In this note, we will describe some of the main known results on residual intersections. We will also add various remarks and list several open problems.

The concept of residual intersection was introduced by Artin and Nagata ([1]). It generalizes the notion of linkage to the case where the two "linked" ideals are not required to have the same height. We will use the following definition from [8]:

Definition 1. Let I be an ideal in a Noetherian ring R and let s be an integer with $s \geq ht\, I$.

a) An *s-residual intersection* of I is an R-ideal J such that $ht\, J \geq s$ and $J = \mathfrak{a} : I$ for some s-generated R-ideal $\mathfrak{a}$ properly contained in I.

b) A *geometric s-residual intersection* of I is an s-residual intersection J of I such that $ht(I + J) \geq s + 1$.

Notice that if I is an unmixed ideal of height g in a Gorenstein ring R, then g-residual intersection of I simply corresponds to linkage and geometric g-residual intersection corresponds to geometric linkage ([15]). Another situation where residual intersection occurs naturally can be described as follows: Let $\mathfrak{a}$ be an ideal in a Noetherian ring R with $s = $ bight $\mathfrak{a} = \max\{\dim R_p \mid p$ minimal prime of $\mathfrak{a}\}$ and assume that $\mathfrak{a}$ can be generated by s elements; now consider a primary decomposition $\mathfrak{a} = q_1 \cap \cdots \cap q_r \cap Q_1 \cap \cdots Q_t$ with $q_i \not\subset \sqrt{Q_j}$ and $ht\, Q_j = s$; then $J = Q_1 \cap \cdots \cap Q_t$ is a geometric s-residual intersection of $I = q_1 \cap \cdots \cap q_r$. It goes without saying however, that many of the technical difficulties in dealing with residual intersections arise in the "embedded" case where the ideal I is contained in some associated prime of the residual intersection J. For further examples we refer to [6] and [9].

---

*Supported in part by the NSF

Cohen-Macaulayness

In their paper [1], Artin and Nagata dealt with the question of when a residual intersection is Cohen-Macaulay. One of their main results however was not quite correct. This was noticed by C. Huneke who was also able to give a correct answer by adding the assumption that the ideal I is strongly Cohen-Macaulay ([6]). Recall that I is said to be *strongly Cohen-Macaulay* if all Koszul homology modules of some (and hence every) generating set of I are Cohen-Macaulay modules ([6]). According to Huneke's result in [6], J is a Cohen-Macaulay ideal of height s, provided that R is a local Cohen-Macaulay ring, J is a geometric s-residual intersection of I, and I is a strongly Cohen-Macaulay R-ideal satisfying G_s. Here, following [1], one says that I satisfies G_s if the number of generators $\mu(I_p)$ is at most $\dim R_p$ for all prime ideals p with $I \subset p$ and $\dim R_p \leq s - 1$ (and I satisfies G_∞ if I is G_s for all s). Later, Herzog, Vasconcelos, and Villarreal replaced the assumption of strong Cohen-Macaulayness by the weaker "sliding depth" condition, but they also showed that this assumption cannot be weakened any further ([5]). On the other hand, the condition G_s can be weakened; this is done in the following result, which also includes the case of non-geometric residual intersections:

Theorem 2 ([8]). *Let R be a local Gorenstein ring, and let I be an R-ideal that is linked in an even number of steps to a strongly Cohen-Macaulay R-ideal K satisfying G_∞ (e.g., let I be an R-ideal in the linkage class of a complete intersection). Consider an s-residual intersection $J = \mathfrak{a} : I$ of I (where as before $\mu(\mathfrak{a}) \leq s$ and $s \geq g = ht\ I$).*

Then J is a Cohen-Macaulay ideal of height s, depth $R/\mathfrak{a} = \dim R - s$, and the canonical module of R/J is the symmetric power $S_{s-g+1}(I/\mathfrak{a})$.

It would be desirable to remove any reference to the G_∞ condition from the above theorem. In fact one should try to prove that in a local Cohen-Macaulay ring R, every s-residual intersection of a strongly Cohen-Macaulay R-ideal I is again Cohen-Macaulay. So far this has been done in the case $s = ht\ I$ ([6]).

For small values of s, the strong Cohen-Macaulay assumption can be weakened: If R is a Gorenstein ring and I is a Cohen-Macaulay R-ideal then every s-residual intersection of I is Cohen-Macaulay for $s = ht\ I$ ([15]); if in addition I is generically a complete intersection and the first Koszul homology $H_1(I)$ is Cohen-Macaulay, then the same conclusion holds for $s \leq ht\ I + 1$ ([11]). This prompts another question: Let R be a Gorenstein ring, and let I be an R-ideal such that I satisfies G_s and $H_i(I)$ are Cohen-Macaulay modules for all $i \leq s - ht\ I$; then is every s-residual intersection Cohen-Macaulay?

Generators and Resolution

Let R be a local Gorenstein ring, let I be a Cohen-Macaulay R-ideal, and let $J = \mathfrak{a} : I$ be a link of I, i.e., a residual intersection with $s = ht\ I$; then $J/\mathfrak{a}$ is the canonical module of R/I ([15]). This allows one to estimate the number of generators of J and - in the graded case - to predict the degrees of the generators. In particular, one can compute residual intersections for $s = ht\ I$, because one has a method of testing whether a given set of elements yields a full generating set of the residual intersection. Furthermore, if I is perfect, then J is also perfect and Ferrand's mapping cone construction provides a means of constructing - at least in principle - a (not necessarily minimal) resolution of R/J from the resolution of R/I ([15]). None of this seems to work if $s > ht\ I$. There are some results concerning the generation of J if $s \leq ht\ I + 1$, and these yield satisfactory answers in case I is Gorenstein ([11]). The generators of J are also known if s is arbitrary, but I is a special type of ideal, namely a perfect ideal of grade 2 ([6]), a perfect Gorenstein ideal of grade 3 ([12]), or a complete intersection ([8]).

As far as resolutions are concerned, it is not even known in the setting of Theorem 2, if a residual intersection of a perfect ideal is again perfect. However if R is regular, then the last Betti number of R/J can be read from the description of the canonical module in Theorem 2 (this is also true for the last graded Betti numbers in the graded case, cf. [13]). Furthermore, the resolution of any residual intersection has been constructed in case I is perfect of grade 2 ([4]), perfect Gorenstein of grade 3 ([12]), or a complete intersection ([2]). If the residual intersection is sufficiently general, then R/J is normal and its divisor class group has been computed ([9]). For each of the above three types of ideals I and for every sufficiently general residual intersection J of I, there is an explicitly known family of complexes that resolve about "half" the modules comprising the divisor class group of R/J ([3], [12], [10]).

It would be desirable to add more special types of ideals I to this list, because even for relatively simple ideals I, the residual intersections and their resolutions tend to provide interesting new classes of examples. Apart from that however, more general information about the resolutions of arbitrary residual intersections is needed, and although an analogue of Ferrand's mapping cone construction may be too much to hope for, it should still be possible to estimate the Betti numbers and predict the degrees in the resolution of a residual intersection.

Codimension

It is interesting to find good upper bounds for the height of colon ideals, since this - among other things - allows one to show by induction on the dimension that two given ideals are equal. In fact it is known that an s-residual intersection is unmixed of height exactly s if I satisfies the assumptions of Theorem 2 (by Theorem 2), if R is Cohen-Macaulay and $s = ht\ I$ (trivial), or if R is Gorenstein, I is Cohen-Macaulay and generically a complete intersection, and $s \leq ht\ I + 1$ ([11]). Furthermore one has the following observation (which was independently noticed by C. Huneke):

Proposition 3. *Let R be a Noetherian local ring which is quasi-unmixed (e.g., let R be a local Cohen-Macaulay ring), assume that $\mathfrak{a} \subset I \not\subset \bar{\mathfrak{a}}$, where $\bar{\mathfrak{a}}$ denotes the integral closure of $\mathfrak{a}$, and let $J = \mathfrak{a} : I$ be an s-residual intersection of I.*
Then $ht\ J = s$, and $\mathfrak{a}$ is generated by s analytically independent elements.

Proof. From [14,4.1] we know that for every $p \in Ass(R/\bar{\mathfrak{a}})$, $\dim R_p = \ell(\mathfrak{a}_p)$, where $\ell(-)$ denotes analytic spread. Thus, since $I \not\subset \bar{\mathfrak{a}}$, there exists a prime p with $\dim R_p = \ell(\mathfrak{a}_p)$ and $I_p \not\subset (\bar{\mathfrak{a}})_p$. In particular $I_p \neq \mathfrak{a}_p$, and hence $J \subset p$. It follows that $s \leq ht\ J \leq \dim R_p = \ell(\mathfrak{a}_p) \leq \ell(\mathfrak{a}) \leq \mu(\mathfrak{a}) \leq s$, which proves our assertion. ∎

Let R be a Noetherian local ring which is quasi-unmixed, and let $J = \mathfrak{a} : I$ be an s-residual intersection. If I is generated by analytically independent elements, then by the above, $ht\ J = s$ and $\mathfrak{a}$ is generated by s analytically independent elements. For the same reason, if I is locally generated by analytically independent elements (e.g., if I is of linear type), then $\dim R_p = s$ for every minimal prime p of J and $\mathfrak{a}$ is locally generated by analytically independent elements. However, is $\mathfrak{a}$ of linear type in case I has this property? This question is known to have an affirmative answer if R is a local Cohen-Macaulay ring and I satisfies G_∞ and the sliding depth condition ([5]) (cf. also [7]). In general, even a negative answer would be quite interesting since a suitable localization of a counterexample $\mathfrak{a}$ might provide a prime ideal in a regular local ring that is not of linear type but nevertheless locally generated by analytically independent elements.

We finish with another observation:

Proposition 4. *Let R be a local Gorenstein ring with infinite residue field, let $\mathfrak{a}$ be a non-Cohen-Macaulay almost complete intersection ideal in R such that $\mathfrak{a}_p$ is a complete intersection for all primes p containing $\mathfrak{a}$ with $\dim R_p = ht\ \mathfrak{a} = g$, let I be the intersection of all the primary components of $\mathfrak{a}$ with height g, and assume that I is a Cohen-Macaulay ideal.*

Then depth $R/\mathfrak{a} = \dim R - g - 1$, and $\mathfrak{a} = I \cap J$ *for some Cohen-Macaulay ideal J of height $g + 1$.*

Proof. We may assume that $\mathfrak{a} = (a_1, \ldots, a_{g+1})$ where $a_1, \ldots, a_g$ form an R-regular sequence and $I_p = (a_1, \ldots, a_g)_p$ for all minimal primes of I. First notice that $(a_1, \ldots, a_g)$ is properly contained in I and set $K = (a_1, \ldots, a_g) : I$. Then by [15], $I \cap K = (a_1, \ldots, a_g)$, K is a Cohen-Macaulay ideal of height g, and $I + K$ is a Cohen-Macaulay ideal of height $g + 1$.

We now show that a_{g+1} is regular on R/K (cf. also [11]). So suppose a_{g+1} is a zero-divisor on R/K, then $\mathfrak{a} \subset p$ for some $p \in Ass(R/K)$. Notice that $\dim R_p = g$ and $I \not\subset p$. Thus $\mathfrak{a}_p \neq I_p$ for some prime p with $\dim R_p = g$, contrary to our choice of I.

Therefore a_{g+1} is regular on R/K, and hence $J = (K, a_{g+1})$ is a Cohen-Macaulay ideal of height $g + 1$. Moreover, $I \cap J = I \cap (K, a_{g+1}) = (I \cap K, a_{g+1}) = (a_1, \ldots, a_g, a_{g+1}) = \mathfrak{a}$.

Finally, $I + J = (I, K, a_{g+1}) = I + K$ is a Cohen-Macaulay ideal of height $g+1$, and thus the exact sequence

$$0 \to R/\mathfrak{a} \to R/I \oplus R/J \to R/I + J \to 0$$

implies that depth $R/\mathfrak{a} \geq \dim R - g - 1$. Therefore depth $R/\mathfrak{a} = \dim R - g - 1$. ∎

By the above, if R is a local Gorenstein ring with infinite residue class field and I is a Cohen-Macaulay R-ideal of height g and positive dimension which is generically a complete intersection; then there exists a $(g + 1)$-generated ideal $\mathfrak{a}$ contained in I such that depth $R/\mathfrak{a} = \dim R - g - 1$ and $\mathfrak{a}_p = I_p$ for all primes p with $\dim R_p \leq g$. This statement is quite close to the original result of [1].

References

[1] M. ARTIN and M. NAGATA, Residual intersection in Cohen-Macaulay rings, J. Math. Kyoto Univ. **12** (1972), 307-323.

[2] W. BRUNS, A. KUSTIN, and M. MILLER, The resolution of the generic residual intersection of a complete intersection, J. Algebra **128** (1990), 214-239.

[3] D. BUCHSBAUM and D. EISENBUD, Remarks on ideals and resolutions, Sympos. Math. XI (1973), 193-204.

[4]   J. EAGON and D. NORTHCOTT, Ideals defined by matrices and a certain complex associated with them, Proc. Roy. Soc. London Ser. A **269** (1962), 188-204.

[5]   J. HERZOG, W. VASCONCELOS, and R. VILLARREAL, Ideals with sliding depth, Nagoya Math. J. **99** (1985), 159-172.

[6]   C. HUNEKE, Strongly Cohen-Macaulay schemes and residual intersections, Trans. Amer. Math. Soc. **277** (1983), 739-763.

[7]   C. HUNEKE, The Koszul homology of an ideal, Advances Math. **56** (1985), 295-318.

[8]   C. HUNEKE and B. ULRICH, Residual intersections, J. reine angew. Math. **390** (1988), 1-20.

[9]   C. HUNEKE and B. ULRICH, Generic residual intersections, in "Commutative Algebra, Proceedings, Salvador 1988", eds.: W. Bruns and A. Simis, Springer Lecture Notes in Mathematics **1430**, 1990, 47-60.

[10] A. KUSTIN, Complexes which arise from a matrix and a vector: resolutions of divisors on certain varieties of complexes, preprint.

[11] A. KUSTIN, M. MILLER, and B. ULRICH, Generating a residual intersection, to appear in J. Algebra.

[12] A. KUSTIN and B. ULRICH, A family of complexes associated to an almost alternating map, with applications to residual intersections, to appear in Mem. Amer. Math. Soc.

[13] A. KUSTIN and B. ULRICH, If the socle fits, to appear in J. Algebra.

[14] S. MCADAM, "Asymptotic prime divisors", Springer Lecture Notes in Mathematics **1023**, 1983.

[15] C. PESKINE and L. SZPIRO, Liaison des variétés algébriques, Invent. Math. **26** (1974), 271-302.

Department of Mathematics
Michigan State University
East Lansing, MI 48824

Some open problems in Invariant Theory

Jerzy Weyman

Department of Mathematics, Northeastern University, Boston MA 02115.

The problems I want to present are connected with the collapsings of homogeneous bundles. Let me start with the general definition.

Let G be a semisimple group and let V be a representation of G. Let us fix a parabolic subgroup P in G. Let us consider a P - submodule M in V, and let N = V/M. Each P-module S defines a homogeneous vector bundle $\mathcal{V}(S) = G \times_P S$ over G/P.

Let us consider the exact sequence

$$0 \longrightarrow \mathcal{V}(M) \longrightarrow \mathcal{V}(V) \longrightarrow \mathcal{V}(N) \longrightarrow 0$$

of vector bundles over G/P. The bundle $\mathcal{V}(V)$ is the trivial bundle V x G/P.

Let $\pi: \mathcal{V}(V) \longrightarrow V$ be the first projection. We define the subvariety $\mathcal{S}(M)$ (the stratum corresponding to M or the collapsing of M) to be the image $\pi(\mathcal{V}(M))$. This notion was introduced by Kempf in [Ke].

Let me list some important examples of strata.

1) Determinantal varieties. Let G = SL(E) x SL(F), V = Hom (E, F) = E*⊗F, G/P = Grass(r, F), M = E*⊗$\mathcal{R}$, where $\mathcal{R}$ is the tautological subbundle on Grass(r, F). Then $\mathcal{S}(M)$ = { $\varphi: E \longrightarrow F$ | rank $\varphi \leq r$ }.

2) Plucker ideals. Let G = SL(F), dim F = n, V = $\wedge^r F$ and G/P = Grass(r, F). Let $0 \longrightarrow \mathcal{R} \longrightarrow F \longrightarrow Q \longrightarrow 0$ denotes the tautological sequence on G/P. We take M = $\wedge^r \mathcal{R}$. Then $\mathcal{S}(M)$ = { $\varphi \in V$ | φ is decomposable }. This is a cone over the Grassmannian Grass (r, F) embedded via Plucker embedding. The defining ideal of this variety is the Plucker ideal.

3) Tensors of a given rank. Let $G = SL(F)$, $V = S_\lambda F$ and $G/P = Grass(r, F)$.

Let $0 \longrightarrow \mathcal{R} \longrightarrow F \longrightarrow Q \longrightarrow 0$ denotes the tautological sequence on

G/P. We take $N = S_\lambda \mathcal{R}$. Then $\mathcal{S}(M)$ is the set of tensors $\varphi \in S_\lambda F$ which have

rank $\leq r$, i.e. they can be written in some basis using only r basis vectors.

4) Nilpotent orbits. For arbitrary G let V be the adjoint representation,

i.e. $V = L(G)$ – the Lie algebra of G. The group G acts on V by conjugations.

Every nilpotent conjugacy class can be realized as a stratum corresponding

to some vector bundle (comp. [K-P]).

5) Strata of the Nullcone. For arbitrary G and V there is natural

stratification of the Nullcone $N(V)$ (the subvariety defined by vanishing of

all G-invariant polynomials on V) by strata which come from the above

construction. In particular all irreducible components of $N(V)$ are of this

type. This is the main result of [H].

6) Multiple roots. In the special case of $G = SL(F)$, $\dim F = 2$, $V = S_n F$. The

space V can be identified with the set of binary forms of degree n. The

subvarieties $X_{p,n} = \{ f \in S_n F \mid f$ has a root of multiplicity $\geq p \}$ are the

strata. The bundle M equals $\mathcal{O}(-p) \otimes S_{n-p} F$, on $G/P = P^1$.

7) Elimination theory ideals (this example was pointed out to me by G.

Kempf). Let $G = SL(F)$, $V = S_{d_1} F \oplus ... \oplus S_{d_m} F$. The space V can be identified

with the set of m-tuples of homogeneous polynomials of degrees $d_1,..., d_m$

in n variables $x_1,..., x_n$. Let us assume that $m \geq n = \dim F$. For $G/P =$

$Grass(n-1, F)$ with the tautological sequence

$0 \longrightarrow \mathcal{R} \longrightarrow F \longrightarrow Q \longrightarrow 0$, let $N = S_{d_1} Q \oplus ... \oplus S_{d_m} Q$,

$M = Ker (V \longrightarrow N)$. Then the stratum $\mathcal{S}(M)$ equals

$\{ (f_1,..., f_m) \mid$ all f_i have a common non-zero root $\}$. The defining ideal of

$\mathscr{S}(M)$ is the ideal we get when we eliminate the variables $x_1, ..., x_n$ from the equations $f_1 = 0, ... , f_m = 0$. In the case of $m = n$ it is the ideal generated by the resultant of $f_1, ..., f_m$.

8) Hyperdiscriminants. Let G be a reductive group, $V = V_\lambda$ - an irreducible representation corresponding to the highest weight λ. Let $Y \subset V$ be the closure of the orbit of the highest weight vector. Let $Y^\vee \subset V^*$ be the dual variety. We define the hyperdiscriminant Δ_λ to be the equation of $Y^\vee$ if codim $Y^\vee = 1$, and a constant if codim $Y^\vee > 1$. This definition is due to Gelfand, Kapranov and Zelevinski ([G-K-Z1, G-K-Z2]). The variety $Y^\vee$ is the collapsing. The variety Y is the cone over $X = G/P$ for some parabolic subgroup P, which is embedded in the projective space by the line bundle corresponding to weight λ. The bundle $N = \xi^*$, where ξ is the natural extension

$$0 \longrightarrow \Theta_X(-1) \longrightarrow \xi \longrightarrow T(X)(-1) \longrightarrow 0.$$

The details of this approach are contained in [W4].

9) The important special case of 8) is when $V = F_1 \otimes ... \otimes F_m$, dim $F_i = f_i$ ($1 \leq i \leq m$), $G = GL(F_1) \times ... \times GL(F_m)$. The space V is the set of m-dimensional matrices of size $f_1 \times ... \times f_m$. In this case the hyperdiscriminant is called the hyperdeterminant.

There are other interesting examples of strata, for example determinantal ideals of symmetric and skew symmetric matrices. The reader should consult [P-W], [J-P-W] for the discussion of these cases.

The basic problem I want to propose is the following.

<u>Problem 1.</u> Find the procedure to determine the Hilbert function and the representation structure of the coordinate ring R(M) of the variety $\mathscr{S}(M)$.

This problem is solved in principle when $\mathcal{S}(M)$ has rational singularities. In this case $R(M) = \bigoplus_{i \geq 0} H^0(G/P, S_i M^*)$ (comp. [Ke], [W1]).

In many cases the cohomology groups can be calculated using Bott's Theorem techniques.

Let me mention what is known about problem 1 for the examples given above.

In the cases 1)-3) above one gets the complete answer, even in characteristic-free way (c.f. [DC-E-P] for example 1, [H-P] for example 2).

Ad 4). In this case the normalisation of $\mathcal{S}(M)$ has rational singularities, so $\mathcal{S}(M)$ has rational singularities if and only if it is normal (comp. [Br], [HI]). As far as the calculation of $R(M)$ is concerned, the bundle M is quite complicated, so Bott's Theorem technique gives an inductive procedure which involves alternating sums (comp. [W1], section 5 for the case $G = SL(n)$, and [Br]). The only combinatorial description of $R(M)$, due to Hesselink (comp. [Br], section 4.4 for the statement) is difficult to use in practice. It is a very interesting problem to find a good, combinatorial description of the multiplicities of irreducible representations in the homogeneous components of $R(M)$. There has been a lot of work done in this direction. Let me mention the work of Gupta and Stanley ([G1, G2, S]).

Ad 6). In this case the problem 1 has been solved in [W2].

For the example 7) the problem is wide open and even special cases would be very interesting.

For the example 8) the problem consists of calculating the degree of the hyperdiscriminant. Both [G-K-Z2] and [W4] contain (essentially equivalent) formulas for this degree. However both involve alternating sums. In this context one has an interesting problem of classifying representations V_λ

for which codim $V^\vee > 1$ (or equivalently deg $\Delta_\lambda = 0$). In the case 9) this problem is solved in [G-K-Z3]. It turns out that in the notation used above, if we assume that $f_1 \geq ... \geq f_m$, then codim $Y^\vee = 1$ if and only if

$$f_1 \leq f_2 + ... + f_m + 2 - m.$$

There are two related problems.

<u>Problem 2.</u> Find the generators of the defining ideal of $\mathcal{B}(M)$.

<u>Problem 3.</u> Find the syzygies of the ring R(M) as a module over the polynomial ring $P(V) = Sym(V*)$.

Let me indicate what is known about the above problems for the examples 1)-7).

For the example 1) the solution of problem 2 is classical (c.f. [DC-E-P] for characteristic-free treatment). Problem 3 was solved by Lascoux [L] in characteristic 0. Kurano [Ku] has shown that the first syzygy is independent of characteristic. It was shown by Hashimoto that the syzygies can be different in positive characteristic (c.f. [Ha], [R-W]).

For the example 2) the solution of 2 is classical (comp.[H-P], [DC-E-P], [A]). The problem 3 is solved for Grass(2, n) in [J-P-W] and for Grass(3, 6) in [P-W]. The general problem can be reduced to calculation of cohomology groups $H^*(G/P, \wedge \cdot N^*)$ but the bundle N^* is complicated, so the calculations become large. However some small cases starting with Grss(3, 7) would be very interesting to have. Let me state one general conjecture about the syzygies of Plucker ideals.

<u>Conjecture.</u> Let us consider the algebra $Tor^{P(V)}(k, R(M))$. Then it is multiplicatively generated by the linear strand, i.e. by

$\oplus_{i \geq 1} \mathrm{Tor}_i(k, R(M))_{i+1}$ (the last subscript denotes the graded component).

For the example 3) both problems are wide open.

Ad 4). The problem 2 was solved for $G = SL(n)$ in [W1]. Problem 3 is solved only in some special cases, for example for the sets of singular nilpotents (comp. [W1], [E-S], [Br]).

Ad 6). There are some partial answers to both questions in [W3].

For the example 7) both problems are open (except for the case of the resultant). Answers to problem 2 would be very interesting even in special cases.

Ad 8). In this case the problem is to find the actual expression for the hyperdiscriminant. This is known in some special cases (cf. [G-K-Z3]).

REFERENCES:

[A] S. Abeasis. On the Plucker relations for the Grassmann varieties. Adv. Math. 36 (1980), 277-282.

[Br] B. Broer. Hilbert Series in Invariant Theory. Thesis, University of Utrecht, 1990.

[DC-E-P] C. DeConcini, D. Eisenbud, C. Procesi. Young diagrams and determinantal varieties. Invent. Math. 56 (1980), 129-165.

[E-S] D. Eisenbud, D. Saltman. Rank varieties of matrices. In "Commutative Algebra", Math. Sci. Res. Institute Publ. NO 15 (M. Hochster, C. Huneke, J. Sally edts.), Springer-Verlag 1989, p.173-212.

[G-K-Z1] I.M. Gelfand, M. Kapranov, A. Zelevinski. Projective duality of varieties and hyperdeterminant. Dokl. Akad. Nauk USSR 305 (1989), 1294-1298; Soviet Math. Dokl. 39 (1989), no 2.

[G-K-Z2] I.M. Gelfand, M. Kapranov, A. Zelevinski. A-discriminants and the Cayley-Koszul complexes. Dokl. Akad. Nauk USSR 307 (1989), 1307-1311.

[G-K-Z3] I.M. Gelfand, M. Kapranov, A. Zelevinski. General discriminants. List of results. Preprint,1989.

[G1] R.K. Gupta. Generalized exponents via Hall-Littlewood symmetric functions. Bull. A.M.S. 16 (1987), 287-291.

[G2] R.K. Gupta. Characters and the q-analog of weight multiplicity. J. London Math. Soc. 36 (1987), 68-76.

[Ha] M. Hashimoto. Determinantal ideals without minimal free resolutions. Nagoya Math J., 118(1990).

[H] W. Hesselink. Desingularisations of varieties of Nullforms. Invent. Math. 55 (1979), 141-163.

[Hi] V. Hinich. On the singularities of Nilpotent Orbits, Preprint 2, February 1990, Dept. of Theoretical Mathematics, Weizmann Institute of Science, Rehovot 76100, Israel.

[H-P] W.Hodge, D.Pedoe. Methods of Algebraic Geometry. Cambridge, University Press, 1952-1954.

[J-P-W] T. Józefiak, P. Pragacz, J. Weyman. Resolutions of determinantal varieties..... Asterisque 87-88 (1981), 109-189.

[Ke] G. Kempf. On the collapsings of homogeneous bundles. Invent. Math. 37 (1976), 229-239.

[K-P1] H. Kraft, C. Procesi. Closures of conjugacy classes of matrices are normal. Invent. Math. 53 (1979), 227-247.

[K-P2] H. Kraft, C. Procesi. On the geometry of conjugacy classes in classical groups. Comment. Math. Helvetici 57 (1982), 539-602.

[Ku] K. Kurano. The first syzygies of determinantal ideals. To appear in J. of Algebra.

[L] A. Lascoux. Syzygies des varieties determinantales. Adv. Math. 30 (1978), 202-237.

[P-W] P. Pragacz, J. Weyman. On the construction of resolutions of determinantal ideals; a survey. In: sem. d'algebre, L.N.M. 1220, 73-92, Springer-Verlag, Berlin (1986).

[R-W] J. Roberts, J. Weyman. A short proof of a theorem of M. Hashimoto. to appear in J. of Algebra (1990).

[S] R. Stanley. The stable behaviour of some characters of SL(n, **C**). Linear and Multilinear Algebra 16 (1984), 3-27.

[W1] J. Weyman. The equations of conjugacy classes of nilpotent matrices. Invent. Math. 98 (1989), 229-245.

[W2] J. Weyman. On the Hilbert functions of multiplicity ideals. Preprint, 1990.

[W3] J. Weyman. The equations of strata of binary forms. J. of Alg. 122 No 1 (1989),244-249.

[W4] J. Weyman. Calculating discriminants by higher direct images. Preprint, 1990.

Baum, R. J., *Philosophy and Mathematics*
ISBN 0-86720-514-2

Eisenbud, D., and Huneke, C., *Free Resolutions in Commutative Algebra and Algebraic Geometry*
ISBN 0-86720-285-8

Epstein, D.B.A., *et al.*, *Word Processing in Groups*
ISBN 0-86720-241-6

Epstein, D.B.A., and Gunn, C., *Supplement to Not Knot*
ISBN 0-86720-297-1

Geometry Center, University of Minnesota, *Not Knot* (VHS video)
ISBN 0-86720-240-8

Gleason, A., *Fundamentals of Abstract Analysis*
ISBN 0-86720-238-6

Harpaz, A., *Relativity Theory: Concepts and Basic Principles*
ISBN 0-86720-220-3

Loomis, L.H., and Sternberg, S., *Advanced Calculus*
ISBN 0-86720-122-2

Protter, M.H., and Protter, P.E., *Calculus, Fourth Edition*
ISBN 0-86720-093-6

Redheffer, R., *Differential Equations: Theory and Applications*
ISBN 0-86720-200-9

Ruskai, M.B., *et al.*, *Wavelets and Their Applications*
ISBN 0-86720-225-4

Serre, J.-P., *Topics in Galois Theory*
ISBN 0-86720-210-6